# Dictionary of Quotations in Geography

To our parents

Emerson F. Wheeler and Ruby R. Wheeler
Alfred M. Sibley and Perla N. Sibley

# Dictionary of Quotations in Geography

Compiled by
JAMES O. WHEELER
and
FRANCIS M. SIBLEY

**Greenwood Press**
New York · Westport, Connecticut · London

**Library of Congress Cataloging-in-Publication Data**

Wheeler, James O.
Dictionary of quotations in geography.

Bibliography: p.
Includes index.
1. Geography—Quotations, maxims, etc.—Dictionaries.
I. Sibley, Francis M. II. Title.
G63.W47 1986 910′.321 85–27243
ISBN 0–313–24196–1 (lib. bdg. : alk. paper)





Library of Congress Catalog Card Number: 85–27243
ISBN: 0–313–24196–1

First published in 1986

Greenwood Press, Inc.
88 Post Road West, Westport, Connecticut 06881

Printed in the United States of America

The paper used in this book complies with the Permanent Paper Standard issued by the National Information Standards Organization (Z39.48–1984).

10 9 8 7 6 5 4 3 2 1



**Copyright Acknowledgments**

Grateful acknowledgment is given for permission to reprint excerpts from the following:

Ellsworth Huntington, "The New Science of Geography," *Bulletin of the American Geographical Society*, 1913, Vol. 45, pp. 641–652. Permission granted by the American Geographical Society.

William J. Berry, "Some Opinions Relative to the Content and Grouping of Geography," *Journal of Geography*, Vol. 32, 1933, pp. 236–242. Permission granted by the National Council for Geographic Education.

W. M. Davis, "Remarks by W. M. Davis on Receiving the Distinguished Service Award of the National Council of Geography Teachers," *Journal of Geography*, 1933, Vol. 32, pp. 91–95. Permission granted by the National Council for Geographic Education.

Elmer W. Ekblaw, "The Attributes of Place," *Journal of Geography*, 1937, Vol. 36, pp. 213–220. Permission granted by the National Council for Geographic Education.

Ellsworth Huntington, "What Next in Geography?" *Journal of Geography*, 1942, Vol. 41, pp. 1–9. Permission from the National Council for Geographic Education.

Harold S. Kemp, "Mussolini: Italy's Geographer-in-Chief," *Journal of Geography*, 1940, Vol. 39, pp. 133–141. Permission from the National Council for Geographic Education.

G. T. Renner, "Geography's Affiliation," *Journal of Geography*, 1926, Vol. 25, pp. 267–272. Permission from the National Council for Geographic Education.

Richard Joel Russell, "Post-War Geography," *Journal of Geography*, 1945, Vol. 44, pp. 301–312. Permission from the National Council for Geographic Education.

E.A. Ackerman, "Geographic Training, Wartime Research, and Immediate Professional Objectives," *Annals*, AAG, Vol. 35, 1945, pp. 121–143. Permission granted by the Association of American Geographers.

Harlan H. Barrows, "Geography as Human Ecology," *Annals*, AAG, Vol. 13, 1923, pp. 1–14. Permission granted by the Association of American Geographers.

Ellsworth Huntington, "Geography and Natural Selection: A Preliminary Study of the Origins and Development of Racial Character," *Annals*, AAG, Vol. 14, 1924, pp. 1–16. Permission from the Association of American Geographers.

Preston E. James, "On the Origin and Persistence of Error in Geography," *Annals*, AAG, Vol. 57, 1967, pp. 1–24. Permission from the Association of American Geographers.

Marvin W. Mikesell, "Tradition and Innovation in Cultural Geography," *Annals*, AAG, Vol. 68, 1978, pp. 1–16. Permission by the Association of American Geographers.

Robert S. Platt, "Reconnaissance in British Guiana, with Comments on Microgeography," *Annals*, AAG, Vol. 29, 1939, pp. 105–126. Permission from the Association of American Geographers.

Carl O. Sauer, "Foreword to Historical Geography," *Annals*, AAG, Vol. 31, 1941, pp. 1–24. Permission from the Association of American Geographers.

Carl O. Sauer, "The Education of a Geographer," *Annals*, AAG, Vol. 46, 1956, pp. 279–299. Permission from the Association of American Geographers.

Edward J. Taaffe, "The Spatial View in Context," *Annals*, AAG, Vol. 64, 1974, pp. 1–16. Permission from the Association of American Geographers.

J. Russell Whitaker, "The Way Lies Open," *Annals*, AAG, Vol. 44, 1954, pp. 231–244. Permission from the Association of American Geographers.

William W. Bunge, "The Geography of Human Survival," *Annals*, AAG, 1973, Vol. 63, pp. 275–295. Permission from the Association of American Geographers.

Charles Redway Dryer, "Genetic Geography: The Development of the Geographic Sense and Concept," *Annals*, AAG, 1920, Vol. 10, pp. 3–16. Permission from the Association of American Geographers.

V. C. Finch, "Written Structures for Presenting the Geography of Regions," *Annals*, AAG, 1934, Vol. 24, pp. 113–120. Permission from the Association of American Geographers.

Norton Ginsburg, "From Colonialism to National Development: Geographical Perspectives on Patterns and Policies," *Annals*, AAG, 1973, Vol. 63, pp. 1–21. Permission from the Association of American Geographers.

Robert Burnett Hall, "The Geographic Region: A Resume," *Annals*, AAG, 1935, Vol. 25, pp. 122–130. Permission from the Association of American Geographers.

Walter M. Kollmorgen, "The Woodsman's Assaults on the Domain of the Cattleman," *Annals*, AAG, 1969, Vol. 59, pp. 215–239. Permission from the Association of American Geographers.

James Parsons, "Geography as Exploration and Discovery," *Annals*, AAG, 1977, Vol. 67, pp. 1–16. Permission from the Association of American Geographers.

John K. Wright, "Terrae Incognitae; The Place of the Imagination in Geography," *Annals*, AAG, 1947, Vol. 37, pp. 1–15. Permission from the Association of American Geographers.

Richard Hartshorne, "The Nature of Geography: A Critical Survey of Current Thought in Light of the Past," *Annals*, AAG, 1939, Vol. 29, pp. 177–658. Permission from the Association of American Geographers.

John Fraser Hart, "The Highest Form of the Geographer's Art," *Annals*, AAG, 1982, Vol. 72, pp. 1–29. Permission from the Association of American Geographers.

Raymond E. Crist, "Geography," *Professional Geographer*, 1969, Vol. 21, pp. 305–307. Permission from the Association of American Geographers.

Barry N. Floyd, "The Pleasures Ahead: A Geographic Meditation," *Professional Geographer*, 1962, Vol. 14, pp. 1–4. Permission from the Association of American Geographers.

Richard Morrill, "The Nature, Unity and Value of Geography," *Professional Geographer*, 1983, Vol. 35, pp. 1–9. Permission from the Association of American Geographers.

Antony R. Orme, "The Need for Physical Geography," *Professional Geographer*, 1980, Vol. 32, pp. 141–148. Permission from the Association of American Geographers.

Francis Delaisi, *Political Myths and Economic Realities*, New York: Viking Press, 1927, pp. 138–140. Permission granted by the publisher.

# CONTENTS

# PREFACE

It has been our great pleasure to stroll through some of the American geography literature of the past four score years or so to select quotations to illustrate the wide range of views on geography and geographic topics. We confined our search largely to the major American periodicals of this century, although a variety of other sources were also tapped. We have ambled through scenes of a rich literature, and we believe that American geography has much to say. We agree with Voltaire (*Irène*, 1788) that "he has invented words which deserve to be quoted." We hope that this collection will be both informative and entertaining.

The quotations are arranged chronologically by topic in order that the reader may gain something of the ebb and flow in the history of geographic thought. Some topics quoted are time specific; others persist throughout the decades. Some quotations are profound and provocative; others are outrageous and pithy. Throughout our searching, we have had Emerson's thought in mind that "next to the originator of a good sentence is the first quoter of it" (*Letters and Social Aims: Quotation and Originality*, 1875). At the same time, we have been tempered in our task by the lines of Edward Young (*Love of Fame*):

> Some for renown, on scraps of learning dote,
> And think they grow immortal as they quote.

The senior compiler has followed the peculiar habit for many years of including a quotation in the upper right-hand corner of class handouts and exams. On the handout listing office hours, most students enjoy reading Mae West's "come up and see me sometime." On the other hand, most students are disapproving of the quotation from Dante on my examination, "Leave every hope, ye who enter!" or Charles Darwin's "we will now discuss in a little more detail the struggle for existence." A number of well-known writers have provided quotations with geographic connotations that the senior compiler has used: Keats, "Are there not other regions than this isle?"; Thoreau, "Live at home like a traveler"; and Shakespeare, "Injurious distance should not stop my way;/For nimble thought can jump both sea and land.'' However, one must proceed cautiously because ''a fine quotation is a diamond on the finger of a man of wit, and a pebble in the hand of a fool'' (Joseph Roux, *Mediations of a Parish Priest*, ca. 1870).

Most of the quotations were gleaned from the major American geographical periodicals of this century, including the *Annals of the Association of American Geographers*; *Bulletin of the American Geographical Society*; *Bulletin of the Geographical Society of Philadelphia*; *Economic Geography*; *Geographical Review*; *Journal of Geography*; and the *Professional Geographer*. In doing so, we have ignored John Milton: ''Borrowing, if it not be bettered by the borrower, is accounted plagiary'' (*Iconoclastes*, XXIII). Nor have we followed Seneca's approach: ''Whatever has been said by anyone is mine'' (*Epistolae*, SVI, vii). Rather, we have taken our quotations faithfully from the originals and have provided citations, even though the authors might not now agree with their own words or wish them repeated. Thus, we have also ignored the warning of Gelett Burgess, who wrote in 1900—five years after he wrote the original ''Purple Cow''—

> Ah, yes, I wrote the ''Purple Cow''—
> I'm sorry, now, I wrote it!
> But I can tell you, anyhow,
> I'll kill you if you quote it.

However, we have been diligent in attempting to maintain accuracy and have kept in mind the words of Winston S. Churchill: ''I am reminded of the professor who, in his declining hours, was asked by his devoted pupils for his final counsel. He replied, 'Verify your quotations' '' (Rudolf Flesch, ed., *The Book of Unusual Quotations* [New York: Harper & Brothers, 1957], p. 232).

The purpose of the *Dictionary of Quotations in Geography* is to review and identify some of the more salient features of American geography. The method is to utilize quotations, many by the most prominent American geographers of this century, to characterize a variety of themes as the discipline has evolved over these past four score years. Something of the changing flavor of American geography emerges from the pens and typewriters of the great variety of authors quoted.

The dictionary is organized into five parts, four of which are based on William Pattison's four traditions in American geography.[1] These are the earth science tradition, the man-land theme, the area studies tradition, and the spatial point of view. The fifth part is entitled ''Other'' and largely deals with geographic techniques, geographic education, applied geography, and the history of geographic thought. In fact, all five parts may be read as the history of geographic thought. Originally, there was to be a section entitled ''Summary,'' but only one appropriate quotation could be found: ''At least a few concluding remarks should be addressed to the woods rather than to the trees. . . . ''[2] In another sense, all the sections represent something of a summary of American geography, though selective and incomplete.

What do Pattison's four traditions mean?

There are four traditions whose identification provides an alternative to the competing monistic definitions that have been the geographer's lot. The resulting pluralistic basis for judgment promises, by full accommodation of what geographers do and by plain-spoken representation thereof, to greatly expedite the task of maintaining an alliance between professional geography and pedagogical geography and at the same time to promote communication with laymen. The . . . traditions [are] in this order: (1) a spatial tradition, (2) an area studies tradition, (3) a man-land tradition and (4) an earth science tradition.[3]

What was Pattison's thesis?

The thesis . . . is that the work of American geographers although not conforming to the restrictions implied by any one of these definitions, has exhibited a broad consistency, and that this essential unity has been attributable to a small number of distinct but affiliated traditions, operant as binders in the minds of members of the profession. These traditions are all of great age and have passed into American geography as parts of a general legacy of Western thought. They are shared today by geographers of other nations.[4]

How do these four traditions differ?

The spatial tradition abstracts certain aspects of reality; area studies is distinguished by a point of view; the man-land tradition dwells upon relationships; but earth science is identifiable through concrete objects.[5]

The sequence of change in American geography was from the earth science tradition to the area studies tradition to the spatial tradition, which has been dominant since the early 1960s. The man-land tradition, contrary to what most American geographers believe, though reaching its peak in the late 1910s and early 1920s, was never so dominant as the rapidly expanding area studies tradition, which quickly replaced the earth science tradition.[6]

Another purpose of this book is to correct a problem stated in 1967 by Preston E. James.[7]

One major source of error that impedes geographic scholarship is the persistent failure of too many geographers to read what other geographers, past and present, have written. The common ignorance of the flow of geographic thought is a surprising characteristic not only of contemporary geography, but also of the geography of centuries past. As a result, the same errors are repeated over and over again, and the theoretical models that seem to illuminate the problems of one generation become the obstacles to the continued development of geographic ideas by the next generation.

Inevitably, we should emphasize geography, like the earth, as having a solid core, though it may have a molten mantle or outer core, which occasionally disrupts at the surface in intemperate rantings, some of which are quoted herein. The compilers of this dictionary agree with Edward J. Taaffe: "Geography is a remarkably diverse discipline, encompassing scholars musing on the beauties of the Laurentian Shield, developing the latest equation for Iowa, or documenting spatial injustice in the city. If the recent history of geographic thought has taught us anything, it should have taught us to accept this diversity and to turn it to good use."[8] And also with John Fraser Hart: "Geography, and geographers, must be permissive, not prescriptive. We should welcome all who desire to be members of our clan, and encourage each one to contribute to the field as best he may, in whatever direction his interests and abilities."[9]

We should also heed the comment of Larry R. Ford: "It would seem that in order properly to purge old geographies, we often exaggerate their failures—emphasizing the most bizarre examples of environmental determinism, the most boring examples of regional geography, the most esoteric examples of quantitative models to show why geography must change and become more whatever (read spatial, scientific, environmentally aware, etc.)."[10] And should we forget the past? Ford says no: "In America, we declare to the world periodically that the real geography has just arrived (will the real geography please stand up?). We do not incorporate and digest earlier works, building upon their strengths and learning from their weaknesses, but rather we reject them totally and forget them."[11]

Sometimes geographers overemphasize their importance, as with the geographer quoted in Antoine de Saint-Exupéry's *The Little Prince*: " 'Geographies,' said the geographer, 'are the books which, of all books, are most concerned with matters of consequence. They never become old-fashioned. It is very rarely that a mountain changes its position. It is very rarely that an ocean empties itself of its waters. We write of eternal things.' " Sir Francis Bacon, a sometime geographer, is supposed to have remarked: "Geography is a heavenly study, but an earthly subject."[12]

The compilers of this dictionary read through many shelves of volumes about geography from which quotations were selected. Some offered more usable quotations than others. We obtained no quotations, however, from Mark Twain or Joan of Arc. We therefore understand what Jessie B. Strate meant in his 1924 *Journal of Geography* article entitled "Mark Twain and Geography" when he revealed that "of the dozen or more volumes which I read, from only one did I get no quotation usable; that was 'Personal Recollections of Joan of Arc.' " Other authors, however, were herein quoted extensively.[13]

Any errors remain the responsibility of the compilers. We would greatly appreciate any of these being called to our attention. We would especially welcome suggestions for quotations that we have overlooked, for inclusion in a future volume.

This project began in the Science Library (Carrell 349) at the University of

Georgia, was transferred to the seventh floor of the Library at Indiana University in Bloomington in the fall of 1983, where the senior compiler could be found on most evenings, and was then completed at the University of Georgia. The junior compiler was found on most afternoons, especially during the summer of 1984, proofing and classifying the quotations.

From the handwritten scribbling of the senior compiler, Audrey Hawkins, Department of Geography, University of Georgia, transformed the quotations into typed copy. We thank her for her considerable assistance, as well as her patience. We also thank Ann B. Clark, Office of the Associate Dean, College of Business Administration, University of Georgia, for preparing the final draft of the manuscript. We also wish to thank Keena Lowe for her assistance in proofing the final manuscript. The senior compiler is writing this preface at 395 Rivermont Road in Athens, Georgia, while gazing, from time to time, at the Middle Oconee River as it carries mud to the Atlantic Ocean, 560 feet below. (What a story this little river could tell, if I could somehow quote it.) But now we are ready to let the geographers speak, not all of whom are so quiet as this river.

As the reader goes through the next many pages, it is hoped that he or she will be as the pupil quoted in Jedidiah Morse's *Elements of Geography*, published in 1866:

*Master:* Presuming, my dear Pupil, that you have an inclination to obtain a knowledge of the entertaining and useful Science of Geography . . . I shall devote a few hours to the pleasant taske of instructing you in some of the most useful and entertaining branches of this Science. . . .

*Pupil:* I hope, Sir, I have a due sense of your goodness; and shall very cheerfully, and I trust, profitably, attend your instruction.[14]

James O. Wheeler
Francis M. Sibley

## NOTES

1. William D. Pattison, ''The Four Traditions of Geography,'' *Journal of Geography*, Vol. 63, May, 1964, pp. 211–216.
2. Brian J. L. Berry, ''Recent Studies Concerning the Role of Transportation in the Space Economy,'' *Annals*, AAG, Vol. 49, September, 1959, pp. 328–342; quotation p. 341.
3. Pattison, p. 211.
4. Ibid.
5. Ibid., p. 215.
6. James O. Wheeler, ''Notes on the Rise of the Area Studies Tradition in U.S. Geography, 1910–1929,'' *Professional Geographer*, Vol. 38, February, 1986, pp. 53–61.
7. Preston E. James, ''On the Origin and Persistence of Error in Geography,'' *Annals*, AAG, Vol. 57, March, 1967, pp. 1–24; quotation p. 2.

8. Edward J. Taaffe, "The Spatial View in Context," *Annals*, AAG, Vol. 64, March, 1974; quotation p. 16.

9. John Fraser Hart, "The Highest Form of the Geographer's Art," *Annals*, AAG, Vol. 72, No. 1, March, 1982, pp. 1–29; quotation p. 5.

10. Larry R. Ford, "Beware of New Geographies," *Professional Geographer*, Vol. 34, No. 2, May, 1982, pp. 131–135; quotation p. 131.

11. Ibid., p. 132.

12. Antoine de Saint-Exupéry, *The Little Prince*, New York: Harcourt, Brace & World, 1943; quotation p. 54.

13. Jessie B. Strate, "Mark Twain and Geography," *Journal of Geography*, Vol. 23, 1924, pp. 81–92; quotation p. 81.

14. Jedidiah Morse, *Elements of Geography*, Boston: I. Thomas and E. T. Andrews, Social Edition, 1876, p. 1.

PART ONE

# Earth Science Tradition

## PHYSIOGRAPHY-GEOMORPHOLOGY

Rollin D. Salisbury, 1910, p. 58. It is true that Physiography is not a success in the hands of every teacher who knows his subject well, and the same could be said of every other subject. A teacher who has not succeeded with Physical Geography is prone to charge the failure to the nature of the subject, but obviously another interpretation is possible.

Frederick William Simons, 1912, p. 277. In order to understand fully the geographic influences which have prevailed in the development of Texas, it becomes necessary, first of all, briefly to treat of its salient physiographic features.

Albert Perry Brigham, 1914, p. 25. In the history of the science of physiography four score or a hundred years carries us back to ancient times.

Nevin M. Fenneman, 1916, p. 23. The physiographic history of the United States may be thought of as a great number of events, constructional and destructional, each affecting a certain area, large or small; of uniform or diverse structure; in general, no two events affecting exactly the same area, but overlapping, some much, some little. Within the area affected by one event, the topography will differ in different parts; it may be because of differing structure or because of the complicating effects of other events.

John K. Wright, 1925, p. 198. America's distinctive contribution to geography is, perhaps, the interpretation of land forms.

Roderick Peattie, 1931, p. 415. There is a question whether or not we should use the term zones in mountain geography. True, there are cases where zones are relatively distinct.

M. R. Campbell, 1932, p. 50. A composite peneplain, as indicated by its name, is one that is not simple in its mode of formation or in its surface features, but is the result of a number of cycles or partial cycles of erosion which are separated by such slight uplifts that the forms of one cycle merge with the forms of the preceding cycle and also with those of the cycle following.

W. M. Davis, 1933, p. 92. It is truly heart-warming for an old teacher to find that he is still remembered by many younger ones. But it must be added that a lively feeling of surprise is mingled with the gratification, because geography as taught in our schools for the last thirty years has not been markedly of the nature that I strove to make it; indeed, the special phase of the old subject which I had recommended as worthy of your attention and which some of you—or your predecessors—had for a time taken up with enthusiasm, has been practically abandoned. In other words, the subdivision of geography formerly called physical geography and later physiography, to which most of my geographical work has been directed and which had something of a "run" in the '90s, has been almost completely replaced by another phase, known as commercial geography, with which I have had nothing to do.

Isaiah Bowman, 1934, p. 177. One cannot speak of the "loss of Professor William Morris Davis." He once said that he would like to return to earth a hundred years after his death to see how much of "the Triassic Formation of Connecticut" had stood the test of a century's criticism.

George D. Hubbard, 1940, p. 60. Penck's goal in morphologic analysis is to interpret diastrophism. Davis' goal was to study geomorphology for its own sake and to discover the basis for geography proper as well as to find much interpretation for geologic history. Davis thought of geomorphology as a part of Geography, Penck as a border field between geology and geography.

Penck depends largely on observation; Davis as much on observation but uses much more theory and reasoning. They differ much in their emphasis on different factors. Davis' contribution is the cycle in all its aspects, while Penck's contribution is in the interpretation of forces and manner of crustal movement. The American student will do well to consider both objectives and both methods so as to keep his own balance in his geomorphic studies.

Howard A. Meyerhoff, 1940, p. 66. The peneplain concept is one of American geomorphology's most serious liabilities.

Carl O. Sauer, 1941, p. 18. Geographers have given strangely little attention to man as a geomorphologic agent.

Arthur N. Strahler, 1949, p. 65. Application of quantitative methods of the study of landforms and the processes which make them is a relatively new field and one which must be developed extensively if geomorphology is to attain a scientific level comparable to that reached by most other branches of natural science.

Richard Joel Russell, 1949, p. 4. The distinction between geological and geographical geomorphology lies chiefly in a contrast between conclusions of vertical and horizontal significance.

Lawrence Martin, 1950, p. 172. A William Morris Davis centenary is a major event in American geography, geology, and meteorology. Evidence does not exist, of course, but no one can deny that the infant Davis's first word in 1850 was probably, *"WHY?"*

Henri Baulig, 1950, p. 188. Geomorphology should ever be thankful to Davis for having freed it from the bondage of classical geology.

G. H. Dury, 1959, p. 2. Geographers as a group deal with man in relation to his environment. As specialists, geomorphologists examine certain aspects of that environment, without considering man at all.

Geoffrey Robinson, 1963, p. 13. Geomorphology, the study of earth sculpture, may be engaged in as a science in its own right.

Geoffrey Robinson, 1963, p. 15. In all sciences the language used for accurate description is mathematics, and mathematical and statistical analyses have been applied to a wide range of geographical and geomorphological problems. It has long been an accepted maxim in science that information, where possible, should be expressed mathematically for the sake of precision. This can be applied to geography. Precise expression, in mathematical terms, with no shades of meaning open to misinterpretation, is infinitely preferable to loosely applied, vague adjectives, which often differ in meaning from user to user and certainly from time to time, and which suffer badly from translation and dissemination in other countries.

Edwin H. Hammond, 1964, p. 12. Land surface configuration is simply a three-dimensional geometry, to which some consideration of surface material is usually added.

Norman J. W. Thrower and Ronald U. Cooke, 1968, p. 181. Slope of the land is one of the most important attributes of landscape to geographers, geologists, land planners, engineers and many other field scientists, yet it is one of the most difficult to measure with precision.

George Harry Dury, 1972, p. 200. If we so define geographical geomorphology, or for that matter geography as a whole, so as to exclude contributions from neighboring areas, we shall invite disaster. Geography, long under attack as what people call a mode of behavior rather than as what they call a discipline in its own right, has everything to lose. Neither grandiose claims nor restrictive limitations will get us anywhere. At this very time, when academic respectability is within our sight and grasp—probably for the first time since, centuries back now, British monarchs could appoint a Geographer to the King—we surely need to demonstrate ourselves capable of interdisciplinary research.

Richard J. Pike, 1974, p. 258. Geographers who become involved with the Moon and Mars and other planets, such as Mercury and Venus, that may come under scrutiny in the near future will face novel and challenging, but by no means alien, problems in landform interpretation. The events and processes shaping the lunar and martian surfaces are not necessarily new or mysterious.

Eric H. Brown, 1975, p. 35. It is observed that at the present time physical geography is internally unbalanced because geomorphology plays too dominant a role. Furthermore, an integrated physical geography has been rediscovered, largely by non-geographers, under the title "environmental science." It is questioned whether human geography any longer need pay heed to the physical world except in a general and elementary sense and whether or not there is a danger of physical geography ceasing to be an integral part of geography and its content being taken over by environmental science.

William L. Graf et al., 1980, p. 279. The eighties will be an exciting time for the geomorphic sciences, especially as viewed from the perspective of the geographer.

## DESCRIPTIVE GEOMORPHOLOGY

C. E. Dutton, 1880–1881, p. 51. Standing upon any elevated spot where the radius of vision reaches out fifty or a hundred miles, the observer beholds a strange spectacle. The most conspicuous objects are the lofty and brilliantly colored cliffs. They stretch their tortuous courses across the land in all directions, yet not without system; here throwing out a great promontory, there receding in a deep bay, and continuing on and on until they sink below the horizon or swing behind some loftier mass or fade out in the distant haze. Each cliff marks the boundary of a geographical terrace, and marks also the termination of some geological series of strata, the edges of which are exposed like courses of masonry in the scarp-walls of the palisades. In the distance may be seen the spectacle of cliff rising above and beyond cliff, like a colossal stairway leading from the torrid plains below to the domain of the clouds above.

Richard Elwood Dodge, 1910, p. 184. The great poets of the world have often attempted to depict in appealing phrases, more graphic than any of the carefully selected and more detailed descriptions of the scientists, the striking features of the landscape that everyone knows but perhaps does not appreciate.

R. H. Whitbeck, 1913, p. 67. Pass now to the area covered by the latest drift. At once a change occurs. The valleys are filled or half filled with mounds and hills. There is no uniformity in the details of the landscape. The symmetrically branching valleys which characterize the driftless area in the limestone region are nowhere in evidence. Streams wind about in confusion. Headwater divides are difficult to locate. There is no system to the stream course. A branch stream may flow nearly north to join a trunk stream that flows nearly south. Large areas have no drainage, and swamps and lakes are met with on all sides. The one impressive fact is that everything on the surface is disordered and yet to it all there is a rounded and graceful contour.

Albert Perry Brigham, 1914, p. 25. In 1795, the brilliant grandson of Jonathan Edwards, Timothy Dwight, who had learned the alphabet at a lesson and read his Bible at the age of four, was elected president of Yale College. Like some other college presidents, he found much to do and found also that he could not endure too much confined and studious life, and he took to travel, kept a journal, and this fruit of his vacations may be found in four volumes, which in their keen and vigorous descriptions of things as he saw them are probably more fresh today than the volumes of theology for which he was famous.

W. M. Davis, 1914, p. 561. If a geographer or a geologist, who has never seen any coral reefs, nevertheless has occasion to give some account of their origin, two courses are open to him. He may simply quote, without expressing his own opinion, the observations and opinions of those who have seen coral reefs; or he may attempt, after a more or less critical study of the reports of original observers, to form an opinion of his own.

Charles Redway Dryer, 1920, p. 4. The earth is a ball filled inside with dirt and worms and covered all over on the outside with nothing but geography.

W. M. Davis, 1933, p. 94. It may be fairly said that, of all outdoor sciences, modernized physical geography is our most constant companion and one of our most enjoyable companions as we travel about; and in these modern days nearly every one travels. The reason for this is that modernized physical geography adds meaning to visible landscapes and thus gives us a better appreciation of the world we live in.

Will F. Thompson, 1964, p. 6. There can be no question that mountains are of scientific interest, and that many studies can be done in such terrain which cannot be done, or done as well, elsewhere.

## EARTH-SUN RELATIONS

Wallis E. Johnson, 1911, p. 68. It is a poor watch . . . that gains or loses as much time in a day as the sun ordinarily does. Not only is the time of sunrise and sunset constantly varying, but the time elapsing from sun noon to sun noon again is sometimes more and sometimes less than twenty-four hours. That the sun is sometimes fast and sometimes slow, at times as much as sixty minutes, . . . must be borne in mind if one would set his watch by the world's great timekeeper.

John K. Wright, 1925, p. 199. Man's craving to ascertain the nature of God and of matter has also molded geographical thought. Geography owes a debt to theology in so far as the "Queen of the Sciences" was one of the few intellectual pursuits in an age of ignorance and tended thus to keep the mind awake and keen. But geography also bears a grudge against theology for the many restrictions that the latter has imposed on the free growth of science in medieval and modern times. Belief in the sphericity of the earth by some medieval theologians was regarded as heresy because certain passages in the Bible may be interpreted to indicate that the earth is flat. Even those in the Middle Ages—and there were not a few—who were ready to concede that the earth is a globe, were none the less inhibited against beliefs in the existence of antipodal inhabitants. How could dwellers on the other side of the earth see Christ at his second coming?

W. Elmer Ekblaw, 1937, p. 215. It might be well to point out that the ultimate control of climate is extra-terrestrial.

A. K. Lobeck, 1942, p. 134. It was once thought that the earth ''sloped'' from the Equator to the poles, and so had different zones, or ''climates.'' *Clima* means a ''slope,'' a root which appears again with a similar meaning in the word ''climax,'' the end of the slope.

S. W. Boggs, 1946, p. 83. If the earth were as large as Jupiter, with a surface area 120 times that of our own planet, it would be much more difficult to think and plan in global dimensions, even in these days of radio and airplanes. But as it is so small it behooves us to think in planetary terms.

Douglas K. Fleming, 1984, p. 76. Neither the logic of ancient Greek philosophers nor the centuries of modern science have permanently dispelled illusions of a flat earth.

## CLIMATOLOGY

V. Stefansson, 1913, p. 19. On a certain Thanksgiving Day which most Dakotans of that time still remember, I followed that wire to the barn. Several times on the way I had to stop and hold both hands over my nose and mouth, for I could not draw my breath otherwise; I did this stooping over the wire, for I did not dare let go of it. I walked with both eyes tightly closed and felt ahead with my feet each time I stepped so I should not collide too forcibly with the barn. With automobile goggles over the eyes I could doubtless have caught glimpses of the barn now and then after getting within fifteen feet or so of it, but though I stopped and opened my eyes momentarily every four or five steps I did not see the wall till my toe touched it.

Robert De C. Ward, 1914, p. 7. With the passing of summer into autumn, the sun's rays become more and more oblique, and weaken; the temperature falls; the days grow shorter; the control of our weather passes gradually but irregularly from the sun back again to the cyclone; from the mildness of summer to the turbulence of winter.

Oliver L. Fassig, 1916, p. 125. The instrument is a greatly simplified form of revolving cloud camera. . . . The camera takes in the entire dome of the sky, from horizon to zenith, at a single exposure of a second or less. The practical use of this camera in cloud topography has been demonstrated; its broader application in geographic, and especially military, exploration is considered.

Robert De C. Ward, 1930, p. 273. A continuous round-the-world voyage, westward from New York to New York, has recently been made by the writer. The trip was planned primarily in order that firsthand knowledge might be gained of the great wind belts over the oceans and the weather types associated with them.

M. H. Shearer, 1930, p. 374. Physical geography is one of the most important sciences needed by young men intending to enter aviation.

Stephen S. Visher, 1930, p. 381. ("The Cyclones.") Cyclones are very numerous and affect mankind significantly each day or two in every locality of the civilized and densely peopled parts of the world.

Robert De C. Ward, 1931, p. 42. Descriptive climatology is more and more becoming anthropo-climatology, just as geography has of late years increasingly become anthropo-geography.

C. Warren Thornthwaite, 1931, p. 655. It is believed . . . the climatic elements of the landscape are here analyzed with a greater precision than has heretofore been attained.

J. E. Church, 1933, p. 529. Snow surveying is one of the newest sciences. Like most inevitable things, it has had several births though few survivals. It was born in Europe, evidently in the nineties, in the study of the density of snow.

Stephen S. Visher, 1935, p. 68. Since Jesus stated, "the Wind bloweth where it listeth," much progress has been made toward an understanding of the factors affecting the weather.

Stephen S. Visher, 1938, p. 627. ("Rainfall-Intensity.") Rainfall intensity is the time distribution of rainfall, or how the rain falls as regards periods of time.

Preston E. James, 1939, p. 132. The extension of regular commercial air lines in Brazil, as in other parts of the world, has been accompanied by an important increase of meteorological knowledge.

Stephen S. Visher, 1940, p. 77. Crop yields are affected rather consistently by specified weather conditions.

Stephen S. Visher, 1943, p. 221. The significance of regional contrasts in precipitation is so great that it is highly desirable that all teachers of geography be well informed thereon.

Charles F. Brooks, 1948, p. 153. Thomas Jefferson was one of the pioneer climatologists of America. He, at Monticello, and James Madison, at Williamsburg, made simultaneous observations of temperature, rainfall, pressure, and wind. Jefferson made five-year averages of temperature, rainfall, and wind from Madison's records to compare with his own, to obtain the differences between coast and interior. Although the SW wind prevailed at both locations, the NE was next in frequency on the coast and the NW in the interior. Jefferson then describes the relative qualities of the NE and NW winds: "The Northeast wind is loaded with vapor insomuch that the salt makers have found that their crystals would not shoot while it blows; it brings distressing chill, and is heavy and oppressive to the spirits; the Northwest is dry, cooling, elastic, and animating."

Arthur N. Strahler, 1954, p. 6. In the study of climates the empirical approach is essentially that of the application of mathematical statistics to records of the climatic elements . . . with emphasis upon mean values, departures from means, probabilities of occurrence of given values, cyclic phenomena, and trends.

F. Kenneth Hare, 1955, p. 162. The writer suggests that if such an account of the atmospheric circulation over a region is considered a necessary part of its written regional geography, then this account must be written with all the rigor and all the sophistication of a full-scale dynamic climatology. Here, as in all other aspects of physical geography, there is no other half-way house than that of mediocrity.

Clarence E. Koeppe, 1957, p. 6. For fifty years and more, geographers in general have struggled with the idea of regional organization in most of the fields of physical and cultural geography. This idea has found expression in the making of regional maps of soils, agriculture, physiography, climate, industry, natural vegetation, and others. There has probably been no field in which there has been a greater diversity of opinion, purpose and results than in the attempt to portray climate on a regional basis.

Ray Y. Gildea, Jr., 1957, p. 433. If land is defined in terms of space and situation, water is a land resource.

James J. Parsons, 1959, p. 205. Rain gauges mounted under Monterey pine and eucalyptus trees on the crest of the Berkeley Hills, California, at an elevation of 1000 feet above sea level have recorded up to 10 inches of fog drip during a single summer, equivalent to nearly one-half of the average rainfall of the area. In one instance 1.03 inches were recorded during a single night. All of this moisture has been wrung from the stratus that rests on the crest of the hills during much of the summer and early fall. This fog drip is a night-time and early morning phenomenon.

Paul E. Lydolph, 1964, p. 291. Climate is such a limiting factor in the Soviet Union that the progression of severe weather types has become a preoccupation in the minds of the Russian people. Cold and drought are the preponderant climatic obstacles to agricultural production and comfortable living. So significant are they that outstanding occurrences have been tagged with specific names, and in many instances these occurrences are so classic in form that the Russian descriptive terms have been accepted the world over.

James A. Shear, 1965, p. 646. Guesswork concerning the annual temperature regime at the South Pole has ended. The temperature curve proves to be congruent with the curve of insolation, with less than 2° F variation in the mean temperatures of the six winter months.

John R. Borchert, 1968, p. 374. But I would prefer to close with the forward look encouraged by the words of President Kennedy or Secretary Brown. And as we look ahead to the era of remote sensing and a new era of resource management perhaps there is prophecy in the notion expressed in a letter which, in truth, I received just a few days before the beginning of this symposium. The letter somehow found its way from an elderly gentleman in southeastern Minnesota to the state agricultural statistician, then to the geography department at the university, from which I have forwarded it appropriately to the state geologist.

It begins, "Dear Sir: Would you please send me all available information on the *glaciers* in Minnesota from the first instance until they finally came *under control*?" I suppose it is possible that those glaciers could get "out of control" again. They could slip out of Canada and overrun everything from the Mesabi range to the Mayo clinic, or from Lake Placid to Levittown. In that case, we will need a good series of synoptic data from remote sensors. We will also need all of the free flow of information and all of the international cooperation that we can possibly get!

John R. Mather, 1972, p. 137. Applied climatology . . . is the scientific analysis of a statistical collective of individual weather conditions directed toward a useful application for an operational purpose.

John R. Mather, ibid. Both Thomas Jefferson and Alexander von Humbolt called for the collection of systematic climatological records for the purpose of answering specific questions involving agriculture, crop introduction, and soil conservation.

John R. Mather et al., 1980, p. 285. Although it is an old field of scientific endeavor, climatology has until recent years been considered an adjunct of the central concerns of meteorology.

## BIOGEOGRAPHY

A. E. Douglass, 1914, p. 321. In the great northern plateau of Arizona, lying at an average altitude of 6,000 feet above the sea, the higher elevations are covered with forests of yellow pine (*Pinus ponderosa*), a fine timber tree with heavy cylindrical trunk and bushy top. The trees are scattered gracefully over the plains and hills, and, with the remarkable absence of undergrowth, render travel through their midst attractive and delightful. For centuries these magnificent pines have stood there, enduring the vicissitudes of heat and cold, flood and drought. The possibility that they might serve as indices of the climate of the past led the author to begin the investigation of the matter in 1901.

Guy-Harold Smith, 1935, p 272. The configuration of the surface is one of the most obvious features of any landscape—unless obscured by a mantle of vegetation.

Griffith Taylor, 1937, p. 291. It has often been remarked that the best map to aid the geographer in forecasting settlement in a new country is one of the natural vegetation.

W. Elmer Ekblaw, 1937, p. 104. ("Significance of Soils.") Because soils have not the power of organic adaptation, nor the quality of mobility, nor the gift of volition, they differ in their character and distribution only as other factors operate upon them.

Hugh M. Raup, 1942, p. 320. The first plant geographers were the "proto-botanists" who first learned how to use plants for food and shelter.

John K. Wright, 1947, p. 13. All persons know some geography, and I venture to think that many of the animals do also.

Charles F. Bennett, Jr., 1957, p. 168. It is not generally known among American geographers that African elephants have been trained for use as beasts of burden in the Belgian Congo. It seems to be generally believed that the African elephant is too wild and intractable a beast to be suited for taming and training such as has been done with its Asian relative.

John K. Wright, 1963, p. 1. My wild geographers were neither savage nor savages in the customary English senses of those terms (though I believe we now consider it more polite to call people "underdeveloped"). My wild geographers, however, were not even underdeveloped; they were courteous, urbane gentlemen, and it was mainly in their personalities, that they exhibited wildness—the wildness that distinguishes the tiger from our pussy, the peccary from the domestic pig, and the wild from the cultivated rose. Non-wild animals have traded, voluntarily or involuntarily, some of their integrity, individuality, and independence as *animals*, for the creature comforts and social security of pig-sty, hearth, or zoo. Similarly, non-wild geographers have relinquished some of the integrity *of their geography* to meet nongeographical human needs or demands or desires. Thus a geographer lacks wildness when he portrays the world in a textbook as consisting of nothing but sweetness and light, out of fear of offending the Pollyannas among his prospective readers.

D. R. Stoddart, 1966, p. 683. Much of the geographical work of the past hundred years . . . has either explicitly or implicitly taken its inspiration from biology, and in particular from Darwin. Many of the original Darwinians . . . had been actively concerned with geographical exploration, and it was largely facts of geographical distribution in a spatial setting which provided Darwin with the germ of his theory.

E. J. Wilhelm, Jr., 1968, p. 123. Since the early years of this century a new emphasis in biogeography, known as ecological biogeography, has emerged and has received the attention of too few geographers.

Werner H. Terjung and Stella S-F. Louie, 1973, p. 109 Ecologists and climatologists have become increasingly conscious of the basic nature of the flow of energy and mass between physical-environmental systems and plant systems. Whereas most events in the universe lead toward increasing entropy or "disorder" of available energy, life temporarily reverses or halts the second law of thermodynamics by using radiant energy of the sun to construct complex biological molecules.

Werner H. Terjung, 1976, p. 199. Once again, and perhaps more urgently than ever before we should address the question: "What is the purpose of climatology in geography?" Is the inclusion of physical geography in many geography curricula a mere tradition springing from the physical science background of the founders of geography, or does the human geographer really need, or see any relevance in climatology. . . .

Philip J. Gersmehl, 1976, p. 223. Finding examples of unexpected environmental repercussions has become a popular hobby. In one place, fertilizing a crop accidentally stimulates algae blooms and fishkills downstream. In another, pesticides raise pest populations by eliminating natural predators. In a third, fire prevention allows a buildup of fuel that leads to even more disastrous fires. Each action initiates adjustments that ultimately impinge on the origin in an unanticipated way.

F. R. Fosberg, 1976, p. 117. Geography is distinguished from other earth sciences as the discernment and delineation of landscape patterns, interpreting the structures and processes that give rise to them, and developing an understanding of their significance in biological and human terms. Ecology is the study of ecosystems, especially local ones, as contrasted with regional, continental, or global ecosystems, which are better treated as part of geography. Biogeography is a subscience of geography comprising the components of geography that are principally concerned with biological organisms.

Thomas R. Vale and Albert J. Parker, 1980, p. 149. The theme of human alteration of vegetation has a long tradition in geography. Assessments of human impacts on contemporary wild vegetation, especially in landscapes commonly perceived to be natural, will likely continue that tradition, bringing a fuller understanding of the ecological role of human activities and serving the needs of nonacademic society. Because of the difficulty of isolating human influence from other factors, however, the human factor is usually best evaluated along with other possible causes of altered plant covers.

Thomas R. Vale and Albert J. Parker, 1980, p. 157. With continued growth, biogeography may well join with geomorphology and climatology in making physical geography a true geography of environments.

## PHYSICAL GEOGRAPHY

Frank Carney, 1911, p. 1 The customary treatment of the physical in geography usually plunges the student at once far beyond his conceptual depth; the floundering immediately becomes distasteful. A study in land forms, as treated in practically all our textbooks, is a topic to the point. When, however, this phase of Physiography is made secondary to the organic side, in which man is foremost, the question of interest will take care of itself.

Philip Emerson, 1911, p. 20 Oldtime schoolwork in geography dealt with numberless details of location. The physical geography of high school courses has presented the general truths of the development of land forms. All elementary geography should lead through the study of important fields of life relations to an appreciation of the larger laws of geography and the clearer and most significant general truths that characterize different nations, industries, land forms and climates.

Nevin M. Fenneman, 1922, p. 20. But this cultivation of the physical side led to more and more human studies. The first distinctive fact about modern American geography, therefore, is that it began at the extreme physical end. Whether or not American geography can ever be said to develop in a deductive way, it certainly began at the right end for such a development. Attention was centered at first on processes and forms without regard to areal distribution. Had it stopped there it could scarcely be called geography, but the fact that it did not stop there showed its inherent geographic tendency. The great importance of this work to sound geographical development can not be questioned.

Harlan H. Barrows, 1923, p. 1. Geography bore many children, among them astronomy, botany, zoology, geology, meteorology, archaeology, and anthropology.

Harlan H. Barrows, 1923, p. 13. I believe that the age-old subject of geography, though it has lost many specialties, still seeks to cover too much ground, and that it would benefit by frankly relinquishing physiography, climatology, plant ecology, and animal ecology.

W. M. Davis, 1933, p. 92. The abandonment of modernized physical geography thirty years ago was inevitable; for as I now see the question, it was introduced prematurely. The subject had not been developed to the point where it could hope to hold a permanent position in our schools. Its explanatory treatment of various problems—and that was its essence—was little understood by superintendents or by the older school principals; it had not been sufficiently studied by the great majority of teachers to make them feel at home in it; and its few texts included too many novelties for easy assimilation. Hence its replacement by some more easily taught phase of geography was unavoidable.

William J. Berry, 1933, p. 237. Geography, of course, like any other subject, may be placed in any combination for administrative convenience. However, if it is to be grouped on the basis of its inherent quality, it would have to be placed among the natural sciences because geography is concerned basically with the earth and not with society.

William J. Berry, 1933, p. 238. As I read writing called social science it appears to be not science but loose, unreasoned, badly informed and illogical. It will be safer at present to put geography with the natural sciences.

William J. Berry, ibid. We feel that geography should be placed with the natural science group for we are convinced that a thoro grounding in weather and climate on the one hand and physiography on the other is necessary for a proper appreciation and understanding of geography.

Brice M. Latham, 1937, p. 193. Within eight years a gully on a farm near Spartanburg, S. C., cut 40 feet into the ground, and canyon-like, knifed its way thru about 1,000 feet of land. For years, no work was done to control this ruinous erosion, and finally the gully reached Bad Lands proportions. More than 40,000 tons of soil were washed from the cut which had been gnawing away at the farm fields thru which it zig-zagged.

Thomas F. Barton, 1939, p. 188. The field of geography has subject matter which offers vital concepts, attitudes, understandings and facts suitable for primary physical science.

Carl O. Sauer, 1941, p. 2. (''Historical Geography.'') Perhaps in future years the period from Barrows' *Geography as Human Ecology* to Hartshorne's late resume will be remembered as that of the Great Retreat. This retraction of lines began by the pulling apart of geography from geology. Geography, of course, owes its academic start in this country to the interest of geologists. Partially in order to gain administrative independence in the universities and colleges, geographers began to seek interests that geologists could not claim to share. In this process, however, American geography gradually ceased to be a part of Earth Science. Many geographers have completely renounced physical geography as a subject of research, if not entirely as one of instruction. There followed the attempt to devise a natural science of the human environment, the relationship being gradually softened from the term ''control'' to ''influence'' or ''adaptation'' or ''adjustment'' and finally to the somewhat liturgical ''response.'' Methodical difficulties in finding such relationships led to a further restriction, to a non-genetic description of the human content of areas, sometimes called chorography, apparently in the hope that by and by such studies would somehow add up to systematic knowledge.

Carl O. Sauer, ibid., p. 5. American geography cannot dissociate itself from the great fields of physical geography. The ways that Davis, Salisbury, and Tarr so clearly marked out must not be abandoned. A geographer, I submit, may properly be a student of physical phenomena without concerning himself with man, but a human geographer has only limited competence who cannot observe as well as interpret the physical data that are involved in his studies of human economies.

Athelstan F. Spilhaus, 1942, p. 431. That the world ocean is a continuous body of water with relatively free interchange between its parts is of fundamental importance to oceanography.

Richard Joel Russell, 1949, p. 2. The drift away from emphasis on physical geography appears to be most characteristic of American geographers.

Glenn T. Trewartha, 1953, p. 84. Much of even physical geography is anthropocentric in character and its materials for study are selected having in mind their importance for the population of the earth.

J. Wreford Watson, 1953, p. 321. Physical geography is of the greatest value to the geographer in making him more exact, more concrete and more disciplined in his observations, analysis and conclusions and thus in helping him to maintain a scientific approach and use a scientific technique in his work.

Arthur N. Strahler, 1954, p. 4. Does the geographer profit more from purely empirical statements of the facts of physical geography or from explanatory descriptions in which the origin of the environmental factors is stressed?

Arthur N. Strahler, ibid., p. 5. The explanatory-descriptive system has its most effective applications in introductory stages of learning about the principles of physical geography.

Arthur N. Strahler, ibid., pp. 7–8. The empirical-quantitative and the explanatory-descriptive methods of physical geography are both valuuable [sic] to the professional geographer and should be strongly represented in his training. The explanatory-descriptive method should be used in introductory geography courses where cultural aims of a general education are paramount.

John Leighly, 1954, p. 224. Geography gained a place in American universities, in the latter part of the nineteenth century, in the form of physical geography.

John Leighly, ibid., pp. 224–225. Within a few years at the beginning of this century Davis's definition [of geography] was narrowed to a concern with the relations of human societies to the earth, and these relations hardened into "influences," directed from nature toward man. This definition attracted to geography a body of adherents whose primary goal was a mechanistic explanation of history. There was a place for physical geography in Davis's conceptual structure, but none in a geography concerned with "influences," in which the physical features of the earth were simply taken as given. Later interpretations of the task of geography have granted to physical geography as subordinate a role as the advocates of "influences" begrudged it. In the hands of those who saw in geography only the investigators of regional economies, it was degraded to an inventory of the physical components of "regions" that have economic value.

David J. M. Hoosen, 1958, p. 269. Russian geography traditionally has been split fairly rigidly into two "scientific disciplines," physical (always dominant) and economic, based on the supposed dichotomy between natural and social phenomena respectively, and exhibiting little cross-fertilization, still less synthesis. Until about the time of Stalin's death, the proportion of the total output devoted to physical geography was consistently about three-quarters. Since then the proportion has declined to about half, and tentative moves have been made towards greater integration through regional studies and changing philosophical orientation within each branch.

Eugene Van Cleef, 1960, p. 8. Until about a half century ago, most geographers considered their field to be an earth science. When discussing either the physical or ontographic aspects of their subject, they viewed them as "earthly." Even determinists, who were inclined to separate man from the planet, pointed out the former's subservience to the physical environment.

Frank Ahnert, 1962, p. 1. For the past few decades, the overwhelming majority of American geographers has devoted its research efforts to the study of problems in human geography, whereas the advancement of physical geography has lagged far behind.

Frank Ahnert, ibid. Worry about the place of physical geography seems to be a specifically American problem rather than a world-wide one.

Frank Ahnert, ibid., p. 4. Thus, the place of physical geography seems identified clearly enough: it stands beside human geography, and is closely connected with it by two-way causal relationships and by the common denominators of location, distribution, and areal differentiation. In this interdependence, physical geography serves human geography, but it also has a goal of its own: it views the physical earth not only as the stage of human affairs, but also as an object of geographical research in itself, with an areal differentiation of phenomena that can be studied quite apart from its possible relevance to the existence and the works of man. This second, independent, goal of physical geography is vital for its other role as a complement to human geography; only if the physical landscape is studied for its own sake can one hope to gain knowledge of heretofore unknown physical-geographical facts and processes that may be later recognized as having an influence on human-geographical phenomena. To study physical geography only from the viewpoint of human geography would have an effect similar to that of studying organic chemistry only from the viewpoint of its application in industry.

B. J. Garnier, 1963, p. 17. Physical geography should concern itself with earth features such as land, soil, plants, air, and water. Each of these is itself studied intensively by one or more of the systematic sciences. Such sciences attempt to understand as fully as possible the objects they study, examining their distribution over the earth and how they have come to possess the characteristics they are known to have. The physical geographer needs this knowledge too, but not as an end in itself. He likes to use it to help him to understand how different phenomena express the particular inter-relationships of particular places, and how a given phenomenon varies in its significance or inter-relationship with other phenomena from place to place.

S. V. Kalesnik, 1964, p. 160. Physical geography is the study of the distinctive characteristics of the state and development of the landscape envelope of the earth. By landscape envelope is meant the *surface* of the globe as the scene of complex and reciprocally interacting atmosphere, hydrosphere, and lithosphere, as the surface on which energy from the sun is received and absorbed, as the locale in which the actions of water, wind, and ice take place with the formation of sedimentary rocks, as the plane on which soil is formed, and as the home of plant and animal life. The landscape envelope is an interrelated system with specific combinations of relief, lithic structure, air masses, water vapor, ocean and lake basins, soils, and living organisms.

William D. Pattison, 1964, p. 215. Only by granting full stature to the earth science tradition can one make sense out of the oft-repeated adage, "Geography is the mother of sciences." This is the tradition that emerged in ancient Greece, most clearly in the work of Aristotle, as a wide-ranging study of natural processes in and near the surface of the earth. This is the tradition that was rejuvenated by Varenius in the 17th century as "Geographia Generalis." This is the tradition that has been subjected to subdivision as the development of science has approached the present day, yielding mineralogy, paleontology, glaciology, meteorology and other specialized fields of learning.

William D. Pattison, ibid., p. 216. The geographer's earth science tradition brings into the open the basis of what is, almost without a doubt, morally the most significant concept in the entire geographic heritage, that of the earth as a unity, the single common habitat of man.

David H. Miller, 1965, p. 1. Whether the surface of the earth is regarded from the cultural, physical or economic side, it is indivisible; geography has an integral focus. To subordinate physical geography into no more than a support for economic or cultural geography violates this unity and allows us to drift into a futile environmental determinism, in which the physical earth becomes only a base to be buried under the more splendid works of man. It also leads us to neglect the earth as a valid object of study.

Preston E. James, 1967, p. 20. In physical geography laws are based on the behavior of vast aggregates of elements. The physicists who deal with the smallest components of matter find that they, too, must fall back on the theory of probability. The human geographers, like the atomic physicists, deal with a relatively small number of individual elements. Actually, then, the two subclasses—physical and human—are not irreducible.

Preston E. James, ibid., p. 21. The distinction between physical and human geography is not a valid dichotomy, and every effort should be made to eliminate it.

Meredith F. Burrill, 1968, p. 5. Davis took *cuesta* from Spanish, gave it a "precise" meaning in English, and then cited features in New Mexico as the archetype. However, in Spanish usage a *cuesta* can be almost any kind of upsticking land form, and relatively few features named "cuesta" conform to Davis' definition.

Melvin G. Marcus, 1979, p. 526. Indeed, an unhappy result of the "environmentalism" dispute was to confuse physical geography with determinism, a leap in logic that mystifies most of us who came later.

Antony R. Orme, 1980, p. 141. DOES physical geography exist? Is it a viable intellectual and professional pursuit with clearly defined goals? Is it needed? My answer to each of these questions is an unqualified "yes." But I have agonized over the questions, wrestled with doubt, and pondered the perceptions of others. To many unbelievers, within and beyond academia, geography is an archaic term linked with tedious rote learning at grade school. Physical geography is seen, if at all, as a subset of that tedium: capes and bays rather than towns and countries. Although practicing physical geographers may deplore these perceptions, regrettably such ideas represent the reality of many persons around us, forcing a defensive posture when offense is needed to meet the challenges of ignorance and misunderstanding. Why is this? The answer lies partly in history, partly in the present benign neglect of the discipline in many areas, and partly in the lack of conviction that characterizes many who are in the best position to demonstrate the need for physical geography to a potentially responsive society.

Antony R. Orme, ibid., p. 148. If physical geography is to survive as a discipline, its practitioners must be proud of its role and its application. During discussions leading to the creation of a new journal in the field, I pondered a number of alternative titles but in the end plumped for *Physical Geography* on the assumption that there was nothing archaic or ambiguous of which to be ashamed, and much to be proud of, among those who study the geography of the physical world. Too often we see physical geography camouflaged in some sexy gown, for example as environmental science, designed to generate fresh interest in the material. No matter how sexy they may be at the time, however, gowns eventually become dated, wear out, and are discarded. Thus the diminishing perception of environmental problems in the face of rising energy considerations has led to a reappraisal of many environmental courses in universities and to some rethinking on the part of society. In the last analysis, therefore, physical geography is as good a term as any, something to be proud of; if we geographers forsake our own terminology in a desire to be different, then God help the discipline!

William Barr, 1983, p. 463. 1983 marks the centennial of a milestone event in the history of scientific research in the polar regions, the First International Polar Year. A total of 14 stations was established in the polar regions by 12 different nations, along with a number of subsidiary stations. Focusing primarily on meteorology, geomagnetism, and auroral studies, scientists at these stations carried out a standardized, synchronized program of observations for a full calendar year.

# PART TWO

# Man-Land Tradition

## RACIAL GEOGRAPHY AND THE TROPICS

William Z. Ripley, 1899, p. 584. Summarizing the views of authorities upon this subject, the almost universal opinion seems to be that true colonization in the tropics by the white race is impossible.

W. M. Davis, 1899, p. 747. Better that the plain [Southern Coastal Plain] should never have grown a pound of cotton, better that its fertile strata should never have emerged from the waters of the sea, than that slavery and its direful, long-lasting consequences should have come upon the United States.

Charles R. Dryer, 1901, p. 390. No people of Ethiopian race has ever risen, without help from some other race, above a condition of barbarism.

Bertha Henderson, 1910, p. 134. [The] study [of geography] is the only rational and sane way to reduce race prejudice. After [one] comes to know men thoroughly he finds but little difference whether they hail from North Germany or Southern India. The sun has burned one a little darker than the other, perhaps the hammer of environment has made one a trifle more energetic and shrewd, but beneath all this, each is swayed by the same emotions, responds to the same appeals, works from the same motives, is in fact "a man for a' that," entitled to respect and consideration and sympathetic helpfulness.

F. V. Emerson, 1911 pp. 15–16. Negro slavery was introduced largely as an experiment and its success even in the South was not assured for nearly fifty years. The negro slave proved far superior to the Indian or the indentured white. He was tractable, capable of hard work, hardy, and easily and cheaply maintained. He could endure heat and malarial conditions. On the other hand he had many inherent disadvantages. He was stupid and incapable of little but simple routine labor. Having learned the care of one crop, it was with difficulty that he could be taught the culture of another. His labor was more or less unwilling, and this, combined with his unintelligence, made necessary a close supervision of his work. Slave labor must be organized and the slaves worked in gangs. It soon came about that the owner of a considerable number of slaves had an economic advantage over the owner of a small number; and out of this fact and the factor of cheap fertile lands, the plantation system developed.

Ellen Churchill Semple, 1911, p. 635. As the tropics have been the cradle of humanity, the temperate zone has been the cradle and school of civilization. Here Nature has given much by withholding much. Here man found his birthright, the privilege of the struggle.

Ellsworth Huntington, 1913, p. 109 ("The Shifting of Climatic Zones.") In order to explain the occurrence of such evidences of high civilization in so unfavorable an environment one of two assumptions appears to be necessary; either the ancient people of America were far more capable than any tropical people of to-day and even than any European race which has yet learned to dwell in the tropics, or else the climate was different from that of the present, so that the tropical forest was replaced by tropical jungle.

George J. Miller, 1913, p. 328. The settlement of Michigan was delayed by erroneous notions of its geography. It was commonly believed . . . that the interior was a vast swamp which might well be left to the fur-bearing animals and to the trappers and hunters.

V. E. Shelford, 1913, p. 29. The geographic importance of an environmental factor is determined by its influence upon organisms. Environmental factors influence animal organisms in one of three ways. (a) They may produce death. (b) They may modify structure or behavior. (c) They may stimulate migrating animals and cause them to turn back when an increase or decrease in the factor is encountered.

W. C. Gorgas, 1915, p. 220. When the great valleys of the Amazon and the Congo are occupied by a white population more food will be produced in these regions than is now produced in all the rest of the inhabited world.

J. Russell Smith, 1915, p. 813. After 2,000 years, reviving Rome, attempting to regain an ancient province, throws herself almost in vain against one of the oldest societies on earth—the Bedouin. Rome once ruled the world—all of it that was worth taking, but she never took the Bedouin. He fled into his desert pastures and left the farm lands to the conqueror. Then Rome fell, and the Bedouin who had helped bring it about came back into his own, turning the Roman farms to pastures. And now Rome and her children, the other Romance countries, are taking another turn and are again driving the Bedouin back from the Mediterranean shores. But he is still a Bedouin, and when they drive him he will continue to be a Bedouin as he was in the days of Charlemagne, of Caesar, and of Nebuchadnezzar, and as he probably was in the bronze age, the stone age, the ice age, and in ages that preceded these.

Ellsworth Huntington and Summer W. Cushing, 1919, p. 342. It seems strange that the finest vegetation should be associated with the most backward types of men.

Andrew Balfour, 1923, p. 1329. So far as the race is concerned I am persuaded that the hot and humid tropics are not suited to white colonization and never will be with our present knowledge, even if they are rendered as free from disease as England.

Robert M. Brown, 1924, p. 43. For a long time, I have suspected that such words as "savage" and "uncivilized" are applied carelessly and with reckless abandon, and that the word "civilized" is too easily appropriated; that the categorical classification of races into Mongolian, Caucasian and the like, or the more simplified (and horrified) Black, White, Yellow and Brown is not a function of geography; and that it is only a perverted human trait to classify mankind as Greeks and Barbarians.

Robert M. Brown, ibid. p. 47. Backward people are backward because the struggle against over-mastering physical controls is too severe.

Ellsworth Huntington, 1924, p. 2. I have been led to conclude that the key to many of the most puzzling features in the distribution of human character is found in natural selection. Through its agency the character of a people appears sometimes to change so rapidly that racial character seems to be plastic rather than fixed as is so often supposed. Yet curiously enough natural selection has been almost neglected by geographers, historians, and other students of human activities. As soon as the selective principle is recognized, however, its effects appear to be so clear that it will probably soon take its place as one of the important principles in geography, history, anthropology, and sociology.

R. H. Whitbeck, 1926, p. 11. ("Adjustments to Environments.") But in the last analysis, it is probable that certain regions and peoples are advanced mainly because of the highly favorable environments in which the race has evolved: that their initiative, energy, and intelligence are products of underlying environmental factors operating upon these people for long ages. It is the old question of race versus place; and the geographer will hold that the *masterful race is the product of the place that nourished it.* And so our cycle returns upon itself— the *place makes the race and then the race progressively remakes the place.*

Glenn T. Trewartha, 1926, pp. 36–37. There exist at the present time two distinct schools of thought with regard to the future of the white man in the wet tropics. The first group believes that the ill effects of tropical climate are due to the various concomitants, viz.: disease, proximity of inferior races, intemperance, etc. These factors are admitted to be handicaps to a successful white invasion of the tropics, but probably not insurmountable ones, and will eventually yield before the assaults of scientific discovery and the increasing demands of mankind for tropical products. The second school believes that, with our present knowledge at least, true acclimatization is impossible. If all of the so-called indirect climatic handicaps are overcome, there still remains the direct influence of tropical sun, heat, and humidity, which act upon the white man's body, and more particularly his nervous system, in such a harmful way as to bring about debilitated, neurasthenic condition.

Mark Jefferson, 1929, p. 661. How interesting it would be to accept the lesson of history and admit freely the first half million foreigners, including Mexicans and other Americans, who attained an A grading in such intelligence tests as were used in the Army to sort our potential officers from the mass of recruits. We have been described as receiving the offscouring of Europe and doubtless we do get some. Why not frame our admissions to the United States on a basis of a clean police and health record and a high intelligence test?

Mark Jefferson, ibid., p. 649. The nations fell into four culture groups which were distinctly regional and were called Teutonic, Mediterranean, Levantine, and Oriental from regions of typical development. Canada, the United States, Australia, and New Zealand were Teutonic of culture, like their mother lands; just as Cuba and the Argentine in South America were Mediterranean, like Spain. Japan, adopting Occidental culture for more than half a century, had attained to a Mediterranean grade. Of the three greatest peoples of the East, Japan has frankly adopted Occidental ways, China would like to adopt them if she knew how, and India must acquire them before she can take much part in controlling her own destinies.

Stephen S. Visher, 1930, p. 152. ("Rainfall and Wind Conditions.") During the last two centuries white man has occupied nearly all middle latitudes. The chief exceptions are the parts occupied by the Japanese and Chinese, where nevertheless, Western civilization has penetrated notably. In contrast to this conspicuous success in middle latitudes is comparative failure in the tropics, where there are very few white families and where white man's civilization has made relatively little impression.

J. W. Hoover, 1931, p. 235. The state of Arizona offers a splendid opportunity for comparison of the influence of ethnic origin with that of natural environment in determining traits and characters of Indian populations. Several distinct physiographic and climatic provinces are represented with important subdivisions. The area is also exceptional in the number of distinct native stocks occupying it.

Robert H. Forbes, 1933, p. 230. The native African negro, far from being the stark savage of common imagination, is the possessor of varied though crude cultures to which in some respects civilized peoples are deeply indebted. These cultures, which vary greatly with local environment, though in some details derived from outside sources, are mainly indigenous and are the net result of raw materials locally available, the needs of isolated peoples, and the ingenuity of these peoples in adapting materials at hand to the satisfaction of their living requirements.

A. Grenfell Price, 1934, p. 42. Those nations that are still attempting white settlement in the tropics may well consider the case of Saba Island. The white inhabitants of this tiny member of the Leeward Islands form one of the very few groups of peoples of northwestern European descent that have conducted a long-standing tropical settlement with success.

Charles E. Kellogg, 1937, p. 146. Race is thus a product of the landscape. People may move but a race is rooted to the soil. The "strong race" has come into physical and social equilibrium with its own landscape. The great civilizations which have made their mark in world history have, for the most part, originated in a particular landscape.

Richard Hartshorne, 1938, pp. 279–280. The second most important racial minority in the United States is the Mexican. . . . As the Mexicans are practically all Indians, they are recognized on sight as belonging to a "colored" race.

Desmond Holdridge, 1940, p. 376. With a few notable exceptions the white man has been a general failure as a colonizer and original producer in the tropics.

Ellsworth Huntington, 1942, pp. 3–4. In the long run, the quality of the people is more powerful than natural resources in influencing civilization and history. From the human standpoint the contrasted energy and activity of the Finns and the Amazonians is certainly one of the major differences between Finland and Amazonia. By Amazonians we mean the sedentary population of mixed Portuguese, Indian, and Negro origin in the lowlands of the Amazon basin. The Finns work far harder and probably more intelligently than the Amazonians. This in turn influences a host of other differences in productivity, trade, transportation, the distribution of cities and so forth. Nevertheless, most geographers let the matter pass with the mere statement that the Finns are industrious and efficient, whereas the Amazonians are shiftless and inefficient.

Ellsworth Huntington, ibid. p. 4. The correct attitude for the geographer would seem to be to recognize, as he usually does, that the Finns and Amazonians are different, inwardly as well as outwardly, and then to measure the difference so far as he is able, and discover their causes. We may find perhaps that the average Finn does twice as much work per hour as the average Amazonian, and works three times as many hours in the course of the year. I have no idea what the real figures are, but these will do for illustration.

Ellsworth Huntington, ibid. pp. 4–5. Innate biological traits . . . cause the two groups to be born with a certain amount of difference in physical vigor, mental aptitudes, temperament, and capacities for development. For example, because the standard of living in Amazonia demands only a little work, the average man there does not get so much exercise as the Finn, probably much less than he needs for good health. Thus his bodily vigor is diminished and his inclination to work is lessened.

Richard Joel Russell, 1945, p. 304. To balance the inanimate character of physical geography came the anthropocentric environmentalism of Ellen C. Semple, who successfully preached the gospel of Ratzel to the American educational world. Geography took the tack of "tell me where you live and I'll tell you what you are." Tropical sun darkened the skin of the Negro while overcast skies blonded the hair and blued the eyes of the Swede. The dark-skinned Eskimo was left out of that particular series but appeared elsewhere, at the bone-awl end of a progression that led to the mills of Manchester and Leeds. Geographers gaily sped off into questions of physiology, nutrition, psychology, sociology and other fields, where they ordinarily met with stern rebuffs. Many regarded it as a birthright that geographers might harvest crops they had neither planted nor tended. The most serious result of the environmentalism fad has been a general discrediting of geography. Scientists began to regard the subject as a pseudo-science and social scientists feared it as an undisciplined catch-all that recognizes no academic boundaries.

Robert S. Platt, 1948, p. 352. ("Environmentalism versus Geography.") In the study of races the general correspondence of a certain skin color with a certain grade of civilization may have been convenient for elementary classification; but, after it has been found that color and culture are not tied together biologically and are not correlated in detail, the old concept becomes a stumbling block to be removed.

Ronald J. Horvath, Donald R. Deskins, Jr., and Ann E. Larimore, 1969, p. 137. America is in crisis. Times of crisis are times for reassessment. It is appropriate that the geography profession assess its position on the matter of race relations in America to determine if the profession is making any significant contributions to a solution to the racial dilemma facing America today.

Donald R. Deskins, Jr., 1969, p. 145. Before geographers can address themselves to the question: What research contributions can the geographical profession make that will contribute to a solution of the racial dilemma presently facing America, they have to answer the question: What have geographers done in the past to contribute to the resolution of American racial problems?

William W. Bunge, 1973, p. 335. ("The Geography.") Environmentalism should never have been purged. True, the work was heavily racist and often absurd, but the basic question of the relationship between man and nature cannot be neglected.

## ENVIRONMENT AND CIVILIZATION

Victor Cousin, 1832, pp. 240–241. Yes, gentlemen, give me the map of any country, its configuration, its climate, its waters, its winds, and the whole of its physical geography; give me its natural productions, its flora, its zoology, &c., and I pledge myself to tell you, *à priori*, what will be the quality of man in that country, and what part its inhabitants will act in history;—not accidentally, but necessarily; not at any particular epoch, but in all; in short,—what idea he is called to represent.

Charles R. Dryer, 1901, p. 370. The great centers of civilization have been located upon lowlands which were traversed by large rivers, or were easily accessible from the sea, or both.

Ellen Churchill Semple, 1910, p. 594. The whole civilization of the Kentucky mountains is eloquent to the anthropo-geographer of the influence of physical environment, for nowhere else in the modern times has that progressive Anglo-Saxon race been so long and so completely subjected to retarding conditions; and at no other time could the ensuing result present so startling a contrast to the achievement of the same race elsewhere as in this progressive twentieth century.

Bertha Henderson, 1910, p. 129. It doesn't take half an eye to catch the expressions of interest and surprise as the student realizes for the first time that geography has been a potent and inexorable factor in shaping his destiny; that there is a definite and fixed relation between the geographical conditions of a region and the progress, intellectuality, and influence of the inhabitants.

Ellen Churchill Semple, 1910, p. 561. In one of the most progressive and productive countries of the world, and in that section of the country which has had its civilization and its wealth longest, we find a large area where the people are still living the frontier life of the backwoods, where the civilization is that of the eighteenth century, where the people speak the English of Shakespeare's time, where the large majority of the inhabitants have never seen a steamboat or a railroad, where money is as scarce as in colonial days, and all trade is barter. It is the great upheaved mass of the Southern Appalachians which, with the conserving power of the mountains, has caused these conditions to survive, carrying a bit of the eighteenth century intact over into this strongly contrasted twentieth century, and presenting an anachronism all the more marked because found in the heart of the bustling, money-making, novelty-loving United States. These conditions are to be found throughout the broad belt of the Southern Appalachians, but nowhere in such purity or covering so large an area as in the mountain region of Kentucky.

W. M. Davis, 1911, p. 233. ("Short Studies Abroad.") [A] university professor of history once showed me an outline map of Europe, which he used as a base for practical exercises, and yet on which the low divide on the plains of central Russia between the north-flowing and the south-flowing rivers was represented by a caterpillar mountain range as strong as the Alps or the Caucasus. I cannot help feeling that history must lack something as long as its scenery is so vaguely conceived and so inaccurately represented; and that it will be better taught and better understood, when teachers and pupils alike, in schools and colleges, have a better knowledge of geography.

Ellen Churchill Semple, 1911, p. 1. Man is a product of the earth's surface. This means not merely that he is a child of the earth, dust of her dust; but that the earth has mothered him, fed him, set him tasks, directed his thoughts, confronted him with difficulties that have strengthened his body and sharpened his wits, given him his problem of navigation or irrigation, and at the same time whispered hints for their solutions.

Ellsworth Huntington, 1913, p. 652. ("The New Science of Geography.") In Arabia . . . because the country is a desert, the great movements have all been outward, never inward. When favorable times allowed of increase of population it has come about almost wholly by growth from within, not by migration from without. Thus Arabia has preserved unity of race, speech, and customs, while India has grown more and more diverse. The one has been prepared by nature for a religion of which democracy is the keynote, the other for a religion bound in the fetters of caste. It could scarcely have been otherwise.

Leonard O. Packard, 1913, p. 33. For nearly five hundred years Russia has been making tremendous efforts to secure shoreline. So real has been the need and so keen the realization of it, that practically all the international relations of the Russian Government can be traced to this source. Indeed, it would be difficult to find in the history of nations another case in which the lack of a single physical feature has exerted so controlling an influence as has the need of shoreline in Russian national and international life.

Benj. A. Stevens, 1914, p. 39. Nothing but the appearance of a force which would render the more favored lowland districts less desirable to live in than the mountain districts, could drive the early inhabitants of Britain into the highlands. Such a force finally appeared in the form of invading bands of Celts who were migrating from the mainland of Europe. *With the appearance of this superior race of people began the positive influence of the mountains upon the history of Britain.* Unable to withstand the invader, the Ivernians retreated to the North and to the West leaving their old homes in the lowland areas in the hands of the Celts. This westward and northward retreat continued until finally the original inhabitants were pressed back into the highland areas where, reinforced by natural forces, they were at last able to beat back the enemy and to establish new homes in security. Thus the mountains began to play their part as the preserver of those weaker races which had once occupied the more favorable parts of Britain.

Walter Fewkes, 1914, p. 662. Among the influences that have powerfully affected man in the West Indies are geological features, climate, ocean currents and winds, fauna, and flora. Among geological influences may be mentioned contour and relief, extent of coast lines, stability and distribution of land that can be cultivated, and different kinds of soils or rocks. Climate has affected agriculture more than any other physical environmental condition by determining the animals and plants available for food. Currents and winds are powerful agents in distributing organic life and determining the direction of human migrations on the ocean.

Albert Perry Brigham, 1915, p. 7. The environment of this day and hour is perplexing enough, but environments change; and man exchanges one environment for another. The steady drive of our environment in its daily flux is replaced by the shock of new environment entered in a day or night or gained by long voyages across the sea. The sum of a man's heredity goes out into his new sphere with him.

R. H. Whitbeck, 1915, p. 584. It is an obvious fact of historical geography that large rivers play an important part in the history of the lands through which they pass.

J. Russell Smith, 1916, p. 131. The conflict between England and Germany was traced in part to contrasting geographical conditions. The continental location of Germany, exposing her to invasion on nearly every side, naturally developed fear of her neighbors and this, in turn, bred a respect in her people for any kind of authority that could protect them. Hence their obedience to military organization and individual subordination. England, on the other hand, has been free from any danger without, because of her insularity and her fleet. This has led to greater political and individual freedom, but also to a greater laxity of organization, as evidenced by internal political dissensions and the relative neglect of education and science.

R. H. Whitbeck, 1916, p. 132. The influence of geographic environment is one of the basic questions of scientific geography. At the bottom, it is a biological question,—Can environment so act upon organisms as progressively to modify them and eventually produce wide variations from the parent stock? If so, how is it accomplished? Are differences in environment, and are changes in environmental conditions competent to bring about the present differentiations among groups of plants, animals, and men?

Ellen Churchill Semple, 1916, p. 151 The French conquest of the strongest robber state dealt a serious blow to the Barbary pirates. Shortly before they had been chastised by American, English, and Dutch warships. But what whipped them was steam navigation. Their rugged coasts could not breed mechanics and engineers.

Alison E. Artchinson, 1918, p. 218. The curtain rises. New players now appear. At an entrance far to the north, coming in by the way of the mighty St. Lawrence, enter the French actors while along the eastern coast the English players step into view. Both advance toward the center of the stage and the dramatic phases of the play are on. But what a contrast in the rate of advance. With almost incredible rapidity the French come up the great highway of the northern river and cut through the long stretches to the heart of the continent. How very, very long the English tarry in the narrow strip of country between the coast and the Appalachian mountains and with what great political and economic results that stay is fraught. Only a knowledge of the lakes of Canada, the great abundance of waterways, and the utter lack of highland barrier can completely explain the Canadian French advance, and how much better it is that that knowledge should not be the mere fact that these lakes exist, but an intelligent interest in the long series of geographical events to which their existence is due. A study of why the low Appalachians were for so long a complete barrier to the English, depends again on a knowledge of the topography and history of the Appalachian mountains, the origin and peculiarities of their passes. At every step the early history of the colonists was conditioned by geography. . . . How different in their effect were the rapid, high gradient, non-navigable streams of New England and the sluggish rivers of tidewater Virginia. Every forward step from the Atlantic to the Pacific was modified in some way by the physical features of the continent.

Charles Redway Dryer, 1920, p. 3. The distribution of ideas and emotions is as truly a part of geography as that of rainfall or corn crops. A map showing in shades of color the prevalence of a belief in transmigration, in the infallibility of the Pope, in the divine right of kings, or any other dogma, doctrine or notion, would be a valuable contribution to human geography. The story of the garden of Eden, which Sir William Willcocks has definitely located, including the tree of life, the serpent, the apple and the fall of man, which forms the basis of the current Christian philosophy is as characteristic of the physiographic conditions of Mesopotamia as the date palm. Such a story and such a system could not have originated in northwestern Europe, in eastern North America or in the Amazon basin.

Ellen Churchill Semple, 1922, p. 38. Thus the economic history of the Mediterranean lands can never ignore the factors of climate, relief, and the uniting force of the *Mare Internum*.

Harlan H. Barrows, 1923, pp. 11–12. Historical geography, the geography of the past, helps to show the significance of past geographic conditions in the interpretation of present-day geographic conditions. It provides the key to many environmental relationships that have persisted after the occasion for them has passed. It introduces, so to speak, the "third dimension" into geography.

Friedrich Ratzel, 1923, p. 180. A certain area, its location being unchanged, always transmits the same impulse to states and nations just as a stream enters a turbulent section of its course, or resumes its quiet, even flow at the same location.

Ellsworth Huntington, 1924, p. 1. Geography is primarily the science which describes and maps phenomena of the earth's surface for the purpose of discovering how the distribution of one set of phenomena is related to that of others. The pinnacle of geography is reached when we are able to explain why certain types of human character, certain lines of progress and stages of civilization are localized in various parts of the world.

Mark Jefferson, 1925, p. 39. Malthus, living on the eve of the age of Steam, feared that men would multiply so fast on the earth that some day they would be brought into terrible distress by hunger. Abundant statistics of the hundred and twenty years that followed have not justified his fears. Food has become over-abundant and population shows distinct signs of being checked by another agency that operates independently of food supply. There is little likelihood that the world will ever be really crowded.

G. T. Renner, 1927, pp. 34–35. Human institutions originate from specific needs of man in his environmental relations. Man's religious institutions arise from emotional urges which drive him to compensate for the inadequacies of his environment and the unsatiated desires of his experience. In accordance with these, nearly every race of men have conceived of an ideal environment as a place to live in during after life—a Valhalla, a Nirvana, a Heaven. But differences in environments give rise to different ideal constructions of heaven. The tropical forest man conceives of heaven as a remote village in a land of inactivity, with no heat or mosquitos, and plenty of wives and yams. The Eskimo places heaven in the warm earth and hell in the cold sky. The Comanches conceived of heaven as a prairie full of bison; the Todas, as a land of pastures and dairies. Heaven to the Hebrew was a city on a height, walled off from the desert nomad. The Mohammedan pictures heaven as a delightful, well watered oasis and hell as a hot, scorching place with hot desert winds to breathe and bitter desert water to drink.

Mark Jefferson, 1928, pp. 217–218. Mobility transforms and ennobles peoples. It has always been so. Mobility along the Nile made old Egypt significant. Mobility on the sea distinguished in turn Phoenicians, Venetians, Norsemen, Dutch, and English. Sea-mobile Carthage compelled sedentary Rome to take to the Mediterranean and greatness. In Rome as in mediaeval Europe a class of horse owners came early into prominence and the name of their element of mobility—the horse—still lingers in their titles chevalier, and marshall. The common Spanish word for gentleman, *caballero*, once meant a horseman just as peasant, *peon,* meant a man on foot.

Derwent Whittlesey, 1929, p. 162. Because of its obedience to rule analogous to that governing human organisms, the study of human occupance of area rests on secure foundations if its dynamic character is recognized and adequately accounted for. Spatial extent is always taken for granted as implicit in the geographic craft. In fact that distribution of people and of their activities over the surface of an earth of varied forms, conditions, and resources, is conceded to be the major premise of anthropogeography, human geography, or chorology, as it is variously called. These spatial concepts remain purely descriptive, however, unless they are treated dynamically, i.e., unless the time factor is cognized.

The view of geography as a succession of stages of human occupance establishes the genetics of each stage in terms of its predecessor.

Mary Jo Cowling, 1929, p. 63. As long as coal lasts in England, so long will the dense areas of population remain coincident with the coal fields. And the coal map and the population map will continue to grow as one.

F. A. Carlson, 1930, p. 260. The soil is an environmental element of paramount importance. It is largely responsible for the retardation or facilitation of the world's civilization.

C. B. Fawcett, 1931, p. 111. More than half the people of the New World speak the English language, and derive with it from England much of the traditions, social ideals, laws and modes of thought which constitute their social heritage—one of the two great factors in their development, the other being their own geographical environment.

Louis A. Wolfanger, 1931, p. 330. Flung in bold relief across the base of the Italian peninsula, the Alpine mountains stand a cumbersome barrier to communication between Italy and central Europe. But how easy it is to travel from the Florida peninsula into southeastern North America. The smooth, level plains make the movement of wagon, auto or train almost effortless.

Glenn T. Trewartha, 1932, p. 80. An Indian village, a frontier fur-trading community for nearly a century and a half, a military post under three flags, a bustling river and railroad town of commercial fame in the mid-nineteenth century—each of these successive tenures profited from the river location and the confluence site, and to a degree they all had their origins rooted in these facts of situation. The present town is a dormant community, with the same locational facts which were the *raison d'être* for the earlier forms of settlement now acting to circumscribe its services and handicap its prosperity.

Stephen B. Jones, 1932, p. 357. Historical events have decided that the great interior plain of North America should be politically divided. Statesmen, looking at the orographical map of North America, might prophesy that two nations could not live peacefully, side by side, in such a continent.

W. M. Davis, 1932, p. 213. What but geographical isolation has permitted the evolution of dialects and tongues? What but our inheritance of five terminal sub-divisions in each of our four limbs from a long line of anthropoid ancesters has led to the selection of 20 instead of 19 or 21 as a "score"? What but geographical out-wandering has determined at once the diversity as well as the relationships of the Romance languages?

R. H. Whitbeck, 1933, p. 25. It would be difficult to find a more impressive example of narrow individualism arising, in part, at least, from an insular habitat. It has been relatively easy to unite the provinces of Canada or the states of Australia into single dominions of continental size, but the island peoples of the

British Antilles refuse to be united. It is a case of insular psychology gone mad, yet when any proposal to unite these subunits is made it meets with determined opposition. Each island's slogan seems to be ''Whom God hath put asunder let not man join together.''

A. E. Parkins, 1934, pp. 224–225. Geography is a study of relationships—relationship between the earth and the life that exists upon it. The geographic environment is a mould into which the human race has been poured, and the history of the race has been shaped by that mould. It is also a mould which today is shaping the economic and political life of races and the character of civilization.

Wallace W. Atwood, 1935, p. 6. As the great human dramas have continued, certain influences of environment have persisted. The boldest of the landscape features may have been modified, climatic conditions may have changed, but the actors in the great human dramas could not escape from the planet on which they were located and they could not escape the more fundamental and persistent influences which determined where they could enjoy good health and secure the materials needed in the production of tools and weapons used in defense. Men do not conquer nature. They learn to make better and better adjustments to the natural environment.

Ellsworth Huntington, 1935, p. 45. Certain historians . . . not only recognize the importance of geographical conditions, but actually utilize them. One of the important steps in the future of geography lies in assisting historians in the geographic interpretation of history.

Stephen S. Visher, 1936, p. 30. At the dawn of history most of the people living in Europe were found in the Mediterranean region. Primitive people were unable to extensively clear forests or to till the sod on grasslands, as they had no effective method of cutting down trees or of plowing up sod. Consequently the people living in the densely forested or the sod-covered sections were chiefly hunters or fishermen.

Isaiah Bowman, 1938, p. 3. The relations of early man to his geographical environment were direct and immediate; those of modern man are complex and in many places indirect.

Helmut de Terra, 1939, p. 101. When we seek to evaluate human evolution in geographical terms, no features are of such significance as terraces. In them we see not only indicators of climatic change and earth movement but also a suitable habitat for early man and his animal contemporaries. Terraces, in fact, are the natural burial grounds of uncounted fossil relics and of the most ancient tools of man and a key to much of the prehistoric past.

J. R. Whitaker, 1940, p. 79. One of the major problems of the twentieth century is the shrinkage of the natural resource base of communities, regions, and nations.

Ellsworth Huntington, 1942, p. 6. Geographers alone cannot solve the problem of national character, but they can join with others in its solution, and they can emphasize the regional contrasts which are apt to be neglected by non-geographers. The problem of national character, especially of the part played in it by physical vigor, offers one of the strongest challenges to the geographer. If we knew as much about the causes and regional distribution of people's physical vigor and inclination toward work as we know about the location and operations of coal mines, for example, geography would make a wonderful new step of progress.

Ellsworth Huntington, ibid., p. 8. According to the most carefully corrected statistics Switzerland's index of health and thus of physical vigor is 93 in comparison with 100 for New Zealand, and 92.5 for the whites of the United States. The figure for Japan drops to only 69. This puts Japan half way toward the low level of India, 45, but probably well above China, for which no data are available. The question for the historian is whether the serious lack of physical vigor suggested by so low an index has anything to do with the way in which Japan hesitates to provoke a strong western power such as England, holds its own with Russia where the index of health is practically the same (70), and tries to lord it over less active people such as the Chinese. If the Japanese were as vigorous as the New Zealanders, where would they stand among the nations? The question for the geographer is how far Japan's poor health is correlated with climate, diet, overpopulation and other geographic conditions. The extreme heat and humidity of the summer, with the consequent high rates of illness and death, are certainly a weakening factor in Japan. Over-population and climate together have a disastrous effect upon diet which is another reason for physical weakness.

Griffith Taylor, 1942, p. 65. It seemed to me clear that those anthropomorphic servants of the environment, King Coal, King Frost, and King Drought, were far more important in producing the population-pattern of saturated Europe than were the actions and orders of men, from Charlemagne and Napoleon downward. The widespread changes due to human orders (i.e., "choice") produced *permanent* results only if they were in line with environmental control, but if not, such changes were soon dissipated or annulled.

Ellsworth Huntington, 1943, p. 26. Inventiveness is an excellent indication of mental vigor and initiative. It is especially important in relation to war, because inventive nations are likely to improve their weapons and devise new tricks to overcome their enemies.

Russell H. Fifield, 1944, p. 297. Man's knowledge of the earth has increased while his methods of warfare have changed. The early Egyptians and Babylonians, located among their river valleys, only knew the importance of land power represented by armies; the Cretans and the Athenians, living along the inland sea of the Mediterranean, realized the significance of sea power expressed by navies; later the British, located on the geographical periphery of western Europe, looked to the world ocean for expansion while the Russians, situated in the heartland of the world island, placed their stress on land power; finally the polar states, noting the Great Circle routes across the Arctic in an age of air power, look to the polar skies of the future.

Robert S. Platt, 1948, p. 351. ("Environmentalism Versus Geography.") An environmental interpretation has been attached to history from epoch to epoch, as well as to human life from region to region. And this interpretation has not been completely exterminated by any objection which can be easily stated.

C. E. Cooper, 1948, p. 237. The higher the civilization the higher the latitude.

George H. T. Kimble, 1951, p. 47. Some people are more advantaged by their physical environment than others. It has taken the people of Denmark one thousand years to bring their dairy farms to productivity levels reached by the people of Illinois, Indiana and Wisconsin in one hundred years.

Griffith Taylor, 1951, p. 7. The modern scientific determinist has an entirely different technique and he knows his environment. Thirty years ago I predicted the future settlement-pattern in Australia. At Canberra (in 1948) it was very gratifying to be assured by the various members of the scientific research groups there, that my deductions (based purely on the environment) were completely justified. This aspect of geography is Scientific Determinism.

Anonymous, 1951, p. 233. Few studies are more useful, few more easily attained, and none more universally neglected, than that of geography. The vast importance of this branch of study is sufficiently obvious to any one who takes the least interest in the passing events of the world, or who receives any pleasure in pondering the revolutions of mighty empires, and their consequences, recorded in history.

John Leighly, 1955, p. 314. Environmentalism was not abandoned completely until the late nineteen-twenties, but it was in retreat through most of that decade.

Wagner, Philip L. and Marvin W. Mikesell, 1962, p. 23. . . . any sign of human action in a landscape implies a culture, recalls a history, and demands an ecological interpretation; the history of any people evokes its setting in a landscape, its ecological problems, and its cultural concomitants; and the recognition of a culture calls for the discovery of traces it has left on the earth.

J. Wreford Watson, 1962, p. 125. There is geography behind every history. Every history writes its own geography.

Walter M. Kollmorgen, 1969, p. 218. One of the interesting anomalies of American history is that no sooner had the stockmen begun to appropriate the western grasslands, when there emerged a meteorological doctrine that challenged the grassland or desert ecology and proposed to replace it with a forest ecology. This interesting meteorological doctrine was not native to the West and did not result from breathing the "superior and intoxicating air" of the West. It was an import from the East and, ultimately, from Europe. In large part this doctrine was based on the tree, as it was among trees that many of our immediate forebears evolved their culture.

Karl W. Butzer, 1980, p. 273. There is then every reason to believe that a continuation of present environmental trends will spell an economic crisis of global proportions within the lifetimes of our children and grandchildren. The negative impacts of such cumulative change will affect some countries, such as the two present superpowers, the United States and the Soviet Union, far more than others, creating severe tensions in the global sociopolitical network. At the same time, it will selectively affect certain regions or economic sectors of individual nations, particularly as resources become inadequate to maintain controlled ecosystems, suggesting further confrontations within existing socioeconomic subsystems.

William M. Denevan, 1983, p. 405. I think that most traditional farmers have more available variability for adaptation than does either an Iowa farmer or corn conglomerate. I know of agricultural strategies in Latin America that once supported millions of people but which professors in the University of Wisconsin School of Agriculture never heard of.

Nicholas Helburn, 1982, p. 450. But it is the change in the quality of rural life that stands out. Some of us may be nostalgic for the old woodburning range, the kerosene lamp, and the Saturday night bath in the middle of the kitchen floor, but most of us are glad to be able to flick a switch, plug in a freezer, and not have to use the outdoor privy.

Nicholas Helburn, ibid., p. 453. The other irreversible decision is related. It has to do with extinction. We accept individual mortality reluctantly, and we rejoice in the ability to reproduce ourselves individually. Birth makes death acceptable, for the community continues. The community includes many species living together in enormously complex interactions. We cannot measure the implications of the extinction of a species, much less a whole community. We have already extinguished some. If we cannot postpone the extinction of others we should, for there is no recall.

## ENVIRONMENT AND BEHAVIOR

Richard Elwood Dodge, 1910, p. 186. Wherever we turn we find that the physical conditions are exerting a great influence on the ways of men, whether in the case of the glacier highways across the mountains of Alaska or in the movements of population in the city of New York. Everywhere we find man following the lines of least resistance, as a rule,—the lines of geographical influences.

N. M. Fenneman, 1911, p. 192. The place of [the] ferries is now taken by fine great bridges, a part of man's response to Nature's clear suggestion.

Ruliff S. Holway, 1911, p. 261. To the physiographer San Francisco is noteworthy for its location on a bay occupying a sunken valley situated on a rising coast. To the geographer San Francisco is interesting as the port at the mouth of the only low pass that opens entirely through the central thousand miles of the Coast Range System which rises abruptly along the western shore line of the United States. To the student of human life San Francisco is remarkable for the energy and courage of its cosmopolitan people—cosmopolitan yet not so removed from a certain youthful provincialism of the West but that they still encourage tales of wickedness of their Bohemian ways rather than point to the sturdy but thoroughly unconventional morality that is carrying the city safely onward to its great future.

Mark Jefferson, 1912, p. 106. The house, especially of the poor man, is made of the materials nearest at hand: if among abundant forests, of wood; where the forest is lacking, of stone or brick. Inheritance, however, is also to be reckoned with.

Walter Lefferts, 1913, p. 180. In these days of excessively high prices of meat, when beef on the hoof sells for ten, eleven and even twelve cents a pound, and men unwillingly become more vegetarian, it is desirable that we should inquire as to foreign possibilities of supplying our demand. Only in a rather primitive plains region, it is obvious, can a supply of cheap meat be procured.

W. A. Cannon, 1914, p. 81. The leading factor which makes exploration of the desert at all possible is the camel, the "ship of the desert" in a very real sense, which can travel a week or more, without water and food, making possible exploration where water can be obtained every 200 miles, more or less.

W. J. Sutherland, 1914, pp. 308–309. There is a direct connection between rich soils, generous rainfall, favorable temperature, and a cultured people.

Charles Redway Dryer, 1920, p. 5. An active child spends most of his waking hours exploring his environment, in finding out what there is in it and what use he can make of everything he finds. In his world of the nursery it does not take him long to find the shortest road to the toy box, the picture book and the candy jar. Some spring morning, after he has learned to walk, he is turned out of doors and his world suddenly expands enormously. He plunges into its exploration with more zest than Columbus or Magellan ever knew, and every day makes discoveries more important to him than the gold of Peru or the spices of the Indies.

Ellsworth Huntington, 1922, p. 210. One of the main objects of geography is to discover the relative effects of different physical factors in determining the location of all sorts of phenomena.

Francis S. Hay, 1925, p. 351. The most obvious effect that sheep have upon the region where they live is that of removing the vegetation.

Carl O. Sauer, 1925, p. 341. The natural landscape is being subjected to transformation at the hands of man, the last and for us the most important morphologic factors. By his cultures he makes use of the natural forms, in many cases alters them, in some destroys them.

Ellsworth Huntington, 1926, p. 345. How about the intelligence with which the farms on the two kinds of land are run? This may possibly be indicated by the amount spent for fertilizer. Here at first sight we seem to find a case where the farmers on the poor land go ahead of the others. In 1919 they spent 38 cents per acre on artificial fertilizers, whereas the farmers on the good land spent only 36 cents. But this difference is trivial.

John Wesley Coulter, 1927, p. 605. Probably most geographers will agree that one of the objectives of economic geography is that of helping man to occupy successfully the difficult places of the earth by pointing out the lines along which adjustments should be made if he is to reap the full reward of his endeavor.

Isaiah Bowman, 1935, p. 43. Popular thought about the permanence of agriculture in our semiarid lands swings from one extreme to the other.

W. L. G. Joerg, 1935, p. 177. It cannot be too strongly stated that we do not approach the planning of natural and other American resources in any spirit of defeatism.

Ellsworth Huntington, 1937, p. 110. The maximum number of births occurs in the late winter or early spring at such a time that the young have a maximum chance of survival. Not only the number of births, but also people's vigor throughout life varies in harmony with the month of birth. Those born most closely in harmony with the fundamental animal rhythm live longer than others. They also are more nearly evenly divided according to sex, and they comprise a higher proportion of persons of unusual ability.

John W. Frey, 1940, p. 451. Many national and international problems are the result of an almost universal demand for petroleum products and of peculiar concentrations in the production and probable occurrence of petroleum.

Carl O. Sauer, 1941, p. 7. If we are agreed that human geography is concerned with the areal differentiation of human activities we are at grips at once with the difficulties of environmentalism. Environmental response is the behavior of a given group under a given environment. If we may redefine the old definition of man's relation to his environment as the relation of habit and habitat, it is clear that the habitat is revalued or reinterpreted with every change in habit. There is not general environmental response in the wearing of straw hats. In Chicago they may belong to the summer wardrobe of the well-dressed man. In Mexico they are the distinctive badge of the *peón* in all weather, and the unmodified Indian wears no hat at all. Like every other culture trait, the straw hat depends on the acceptance by a group of an idea or mode, which may be suppressed or substituted by another habit.

J. R. Whitaker, 1943, p. 15. Opportunities for teaching the lesson of responsible stewardship of natural resources lie at every hand in geographical study.

J. R. Whitaker, ibid., p. 17. Perhaps the basic objective of conservation education is to develop an appreciation of the value of natural resources, of our dependence on them, and of the necessity for maintaining adequate supplies of them. A farmer of Laurence County, Tennessee, moved from farm to farm, bragging of the farms he had worn out. Today a son of his is struggling to make a living on one of those worn-out farms.

Isaiah Bowman, 1945, p. 216. The earth is *changing* in relation to man: this is the distinctive contribution to geographical science from Herodotus to Von Humboldt, Vidal de la Blache, and Mackinder. The new geography is new in materials and devices, and in knowledge of the laws of modern science. But its central principle has remained the same for countless generations. Whether men use the advantages of the earth for good or evil ends is a question in social and political morality. Science leaves the field at this point with a single challenging conclusion: the earth is big enough and rich enough for us all if we learn how to live in peace.

George H. T. Kimble, 1951, p. 48. I believe it is a mighty *good* world. In spite of all the man-handling it has had, and all the abuse, it still abounds in noble prospects and physical satisfactions.

Stephen S. Visher, 1952, p. 98. It is much easier . . . to point out how, in Indiana, extensive production of cantaloupes correlates with sandy soil and with the early coming of the summer heat than to establish the conditions that correlate with the production of men in a particular category, distinguished chemists for example.

Marguerite Keinard, 1953, p. 190. Geographical music can be used as the basis for class discussions about how the environments and activities of people are reflected in their music. For example, is not the vast monotony and loneliness of the Russian steppes reflected in the sombre, melancholy quality of much of the music of this area? Are mountain peoples assisted in composing their lively folk tunes by the clear invigorating mountain air? Or perhaps their very isolation is reflected in some of their more wistful melodies. Is the easy rhythm of our Negro spirituals affected by the warm, humid conditions under which the Negro once slaved? Do the wild, frenzied dances of the Slavic peoples show their intense joy at completing a bountiful harvest which assures them a safe winter? Does the languorous music of Hawaii parallel the attitudes of a people whose environment is so kind to them that they can live an easy, carefree life?

Robert D. Campbell, 1968, p. 748. The regional geographer can effectively make use of studies in [other] fields if he can put aside, at least for the time, the difficult problem of the effect of environment upon personality formation.

Walter M. Kollmorgen, 1969, p. 215. The western cattleman was an illegitimate intruder whose professional ancestors were spawned in the grasslands of northern Spain, and as such he was not a welcome progeny in the cultural hearth of Anglo-America.

Walter M. Kollmorgen, ibid. It was a century ago that the American cowboy became conspicuous on the western horizons, accompanying Texas longhorns on their long treks northward to newly founded cattle towns and northern pastures. In the background of the longhorns and the cowboys there soon emerged another image, that of the western cattleman, who appropriated the bounty of grass on a scale and in a manner completely foreign to the ways of the eastern woodsman.

Walter M. Kollmorgen, ibid., p. 238. Our fumbling experiences with our western lands may well prompt us to ask ourselves how much of our geographic impressions come from our mothers' milk, or to what extent they are objectively derived.

James Parsons, 1977, p. 2. The faculties of description and evaluation are those most in need of cultivation if we are to interpret the relationships of land and life and better illuminate the esthetic qualities of landscapes so that men may live more wisely and happily.

Nicholas Helburn, 1977, p. 336. Our conclusion is manifest in truths that seem familiar, that the good life is balanced with some civilization, some wildness, some downtown and some Walden. We all need a change of scene, but some of our frantic escape from the city derives from the one-sidedness of the city. Recognizing the universal need for naturalness in our lives, we can build and care for more harmonious landscapes across the whole continuum.

Marvin W. Mikesell, 1978, p. 3. In any case, studies of deforestation, erosion, and other processes of environmental modification have constituted one of the most persistent themes of cultural-geographic inquiry.

John Fraser Hart, 1982, p. 24. Any geographer who takes leave of his armchair, and gets up and goes out and takes a close look at the world around him, is going to be forced to conclude that Mother Nature still plays a major role in human affairs.

Nicholas Helburn, 1982, p. 447. Geography widens our horizons, helps us become cosmopolitan. Characteristically, an individual's life space is small. A few hectares is all that we can take care of. A few hundred hectares are all we experience in a characteristic day.

Geography helps by providing vicarious experience, for we cannot know everywhere first hand. Further, it helps by giving context within which the vicarious experiences of literature and the mass media make more sense.

Nicholas Helburn, ibid., p. 453. Collectively, we carry an enormous responsibility to institute a permanent agriculture—a responsibility to our children and their children, a responsibility to the hungry and malnourished all around the world.

## CLIMATE AND MAN

Richard Elwood Dodge, 1910, p. 181. It is not by chance that several of the great religions of the world have been the products of the arid regions. The clear skies, the brilliant stars, the far-reaching visions, the wonderful colors of the rocks, have found expression in poetry and legend and man has been led to high things as he has wandered amid the exhilarating silence of the deserts. The high plateaus of Asia Minor to which man came from the eastern deserts and from which he looked out onto the beautiful sea with its softness and changing tones, led to the classic poetry and higher thoughts of reverence and fellow sympathy we find reflected in the Bible. Man looked down from the high places and sought to express in word and action the effects of the wonderful environment on his imagination.

Amid such scenes of unending charm and impressiveness developed the essentials not only of Christianity but of Mohammedanism and Buddhism. In no other part of the world are man's thoughts so naturally led to the higher things as in these vast expanses full of beauty and power that lead to lofty thinking.

Great in contrast to these scenes are the regions of the world of depressing cloudiness and limited views, where all is dull, dreary and uninspiring, where life is hard and the details monotonous, and where man's thoughts, even those of civilized man, are with difficulty brought to spiritual topics.

V. Stefansson, 1913, p. 27. The Eskimos a hundred miles farther north live all winter in houses of wood and earth kept at so high a temperature that most hours of the day the sweat streams from the people that sit around within doors stripped to the waist; when they go out they are dressed in double suits of furs that keep them entirely warm all day; if they run or work hard their under clothes are often wet with perspiration at night.

Ellsworth Huntington, 1913, p. 116 ("The Shifting of Climatic Zones.") In order to test the effect of changes of weather upon mental activity, curves have been plotted showing the variations in speed and accuracy of three persons as measured by daily tests on the typewriter for a year. These curves seem to show the existence of an intimate relationship of some sort between the fluctuations of the barometer and changes in mental ability.

Mark Jefferson, 1913, p. 161. North America, if we include the West Indies and the Arctic islands, is notably more populous in the south than in the north.

J. Russell Smith, 1915, p. 814. Why does this man of the Desert's Edge have such unending continuity while we of Europe and America have such unending change? The answer is simple. He of the desert is adjusted to his environment; we are not. Having become adjusted to his environment, he has small reason for changes, and we, being still unadjusted and, because of inventions, increasingly unadjusted to ours, have need, ever increasing need for change. We live in a growing world. His has been finished for ages.

Ellsworth Huntington, 1916, p. 14. Civilization and climatic energy appear to go hand in hand. This suggests the far-reaching hypothesis that a stimulating climate is an essential condition for civilization.

Ellsworth Huntington and Sumner W. Cushing, 1920, p. 257. Happy are the boys and girls who live on a sugar cane plantation! Nature's stick of candy, a piece of sugar cane, can be had at any time for the cutting. Moreover, it can be chewed and sucked much longer than an ordinary stick of candy, and will still taste delightfully sweet. The boys and girls who would enjoy this happy experience must betake themselves to warm regions.

G. T. Renner, Jr., 1926, p. 272. (''Geography's Affiliations.'') Geographers [have been] asked to avoid having anything to do with the affairs of men which are ugly and sordid but instead devote most of their study to the environment which is pleasant and clean. In other words, the appeal is made for more ''fresh air and sunshine'' in geography. Fresh air and sunshine are indeed essential in geography but they themselves are not constant factors. They are one thing in the small village but quite another to the slum or apartment houses of New York City. They are interwoven very differently into men's lives in Southern California, Southern Chile or Spitzbergen. Even these factors then possess significance only when viewed in relation to the economic and social life of man. A climate is hot or cold or disagreeable or pleasant only when measured in terms of man's reactions. Thus a geography which is not a human geography is indeed a futile thing.

John H. Wigmore, 1929, p. 115. Mohammedan law is in its origin a law of the desert. The close relation between the region of Mohammedan law and the great continuous area of scant rainfall (below 10 inches) and of interior-basin drainage in the Old World is striking. At the same time within the system there are interesting differences for the desert proper and the oases: as between Bedouin Law and the Egyptian Penal Code. On the other hand, climate seems to have had no influence on the Buddhist religious legal system, which, originating in the warm plains of India, spread northwards into the icy plateau of Tibet and the dreary steppes of Mongolia.

J. Wright Baylor, 1931, p. 264. When one finds himself in the midst of a blinding sandstorm on the Central Washington desert as he winds his way thru the purple sage, startling the jackrabbit from his place of refuge on the leeward side of a desert shrub, one cannot fail to appreciate the dominance of climate as a controlling factor in man's activity. In face of such desolation and waste, the few settlers now inhabiting this desert are forced to locate their places of abode, as are the tribes of the Sahara, in relation to the few available sources of water.

Isaiah Bowman, 1935, p. 43. When we deal with large areas of dry land in which are to be anchored whole blocks of population that must take gambler's chances on the rain—and periodically seek public relief—we talk less of conquest and more of compromise with nature.

John Kerr Rose, 1936, p. 88. The correlation of variations in corn yield with fluctuations of climatic factors is more than an academic problem. Many of the human activities prominent in the Corn Belt, rural or urban, are directly or indirectly related to corn. Beginning with the preparation of the seedbed in spring and extending intermittently through the harvesttime of late autumn, the larger part of the Corn Belt farmer's time and activities is applied to corn. Wintertime activities include the feeding of animals with corn. Corn Belt villages exist mainly to serve the corngrower. The cities buy, process, and sell the products of the cornland and serve the corn farmer in many other ways.

H. F. Raup, 1936, p. 264. The American appetite for green foodstuff has grown to demand fresh lettuce, beans, celery, and other vegetables, regardless of weather conditions, as the possibilities of scientific refrigeration and fast transportation have developed.

George F. Deasy, 1937, p. 106. Popular, uninformed opinion in most large cities today supposes that man, other than the farmer, has been largely freed from significant climatic influences. The average urban dweller no longer exhibits that anxious concern about what tomorrow's weather will bring that characterized his forefathers. It is true, no doubt, that he may lament a rain that spoils a Sunday game of golf, or is discomforted in his attempts to sleep during an excessively hot and humid period, but that the weather can and does vitally affect the pocketbook and the life of himself, and his fellowmen is a thought that seldom if ever enters his consciousness.

Raymond H. Wheeler, 1939, p. 99. During warm periods, the culture pattern, as statistically determined by the distribution of these variables, is classical, aristocratic, organic, and totalitarian. As warm periods shift to cold, the culture pattern shifts to its opposite and becomes romantic, democratic, atomistic, proletarian, and individualistic. More specific among the warm-climate variables are rationalism, deduction, idealism, abstract art, socialistic views, cooperation theories of biology and science, evolution, vitalism, teleology, geometry, and epigenesis. Among the cold-climate variables are empiricism, induction, utilitarianism, laissez-faire, anatomy, preformation, algebra, competition theories of biological and social evolution, concrete and imitative art.

Griffith Taylor, 1939, p. 177. In Australia . . . human occupation is very definitely controlled by rainfall.

Eugene Van Cleef, 1940, p. 25. In 1915 I undertook a field investigation of Finnish settlements in northeastern Minnesota, in the hope of determining whether or not these settlements are in any way affected by geographic conditions with which they come into contact. I was not interested especially in the theory of environmentalism, but rather in an opportunity to test out the more or less tenuous theory held at that time, that human migration was effected [*sic*] along isotherms.

Shannon McCune, 1940, p. 66. Tyosen (Korea) is a land of varied geographic landscapes. This diversity is due to the contrasts in the modified continental climate from region to region.

Thomas F. Barton, 1941, p. 226. Weather arouses such a universal interest and curiosity among young and old, rich and poor, and the slow and the quick to learn that a weather station should exist for every school.

Stephen S. Visher, 1941, p. 302. The South has been called America's "problem region." Many southern people are financially poor and culturally below the American standard. Various influences have been blamed for their status. Among these are the Civil War, the Negro, the dominance of cotton, and the inadequacy of local capital. Geographic conditions have also occasionally been blamed, especially the long hot season and the comparative infrequency of stimulating changes of weather.

Rainfall conditions, however, seldom have been considered to be significant in understanding the South. The purpose of this article is to point out some rainfall conditions which clearly adversely affect the South.

Isaiah Bowman, 1945, p. 214. The earth is not a perfect home for man. Some of the drained swamps of Florida, now truck gardens, will not produce acceptably unless copper is added. Most soils are deficient in something, most climates call the tune on man's comfort and crop possibilities.

Earl B. Shaw, 1963, p. 74. An analysis of how . . . these factors influenced this . . . series may be useful to teachers who wish to study geographic influences upon baseball. . . . The influence of weather began with a day's postponement of the fifth game, because of rain.

Kenneth Thompson, 1969, p. 50. Even the name "California" meant earthly paradise. And foremost among California's paradisaical qualities was the vaunted "salubrity" of its environment. Typically in the nineteenth century the word most used to describe California was "salubrious." So great was the emphasis on the salubrity of California that attention seems to have been diverted from the fact that not all of the state was viewed as outstandingly beautiful.

Walter M. Kollmorgen, 1969, p. 217. To put it boldly, we may start by saying that there may be little relationship between geographic perception and geographic reality. We may even argue that geographic perception is more subjective than objective, and that it is always conditioned by experience and inherited values. If this is true, then human action is conditioned not by reality, but what is assumed to be reality. It is in this light that we must view the notion of the Great American Desert, stereotype images of the West generally, and the prospects of converting much or most of the Great American Desert to family farms.

Walter M. Kollmorgen, ibid., p. 218. The belief that weather and climate result in part from an interaction between atmosphere and vegetation was widely accepted in Europe in the eighteenth and nineteenth centuries, and trees played a major role in the realm of the vegetation they knew. Man (and animals) could also play an important role in climate by fostering or destroying certain kinds of vegetation, particularly trees. Thus, Gibbon, in his famous history, took note of the "intense frost, and eternal winter" of early northern Europe, which was subsequently modified by "modern improvements." The latter consisted of clearing the woods, "which intercepted from the earth the rays of the sun," of draining "the morasses," and of "cultivating the soil."

## MAN-LAND GEOGRAPHY

W. M. Davis, 1902, p. 235. It is the relationship between the physical environment and the environed organism, between physiography and ontography (to coin a term) that constitutes the essential principles of geography today.

William Morris Davis, 1906, p. 71. I am disposed to say that any statement is of geographical quality if it contains a reasonable relation between some inorganic element of the earth on which we live, acting as a control, and some fact concerning the existence or growth or behaviour or distribution of the earth's organic inhabitants, serving as a response; more briefly some relation between an element of inorganic control and of organic response. The geographical quality of such a relation is all the more marked if the statement is presented in an explanatory form. There is, indeed, in this idea of a causal or explanatory relationship the most definite, if not the only, unifying principle that I can find in geography.

Charles Redway Dryer, 1912, p. 73. The "new geography" may be defined as the study of the relations of earth features to one another, culminating in the relations of human societies to their natural environments.

W. J. Sutherland, 1914, p. 311. Life responses together with inorganic resources always determine the industries of a region. Hence industry is a secondary response to physical environment.

George J. Miller, 1915, p. 129. Today there is practically a universal agreement of what constitutes scientific geography. The European concept, which has found some followers in this country, emphasizes the distribution idea. The American concept retains this idea but adds something far more vital, viz., the idea of life relationship. Hence modern scientific geography may be considered to be a study of the response of life to the natural environment. This retains everything desirable in the older concepts on the subject, but places emphasis upon the cause and effect idea. It makes possible a sharp cleavage between the science of geography and all the other subjects.

Richard E. Dodge, 1915, p. 305. To know one's landscape, to feel in sympathy with it, is often to be at peace with life. When all the world seems wrong and the burdens overwhelming he can look out on the familiar fields and hills or get among them and give way to their beauties of form and color as a resource within himself that will be an ever present power of recuperation.

*Christian Science Monitor*, 1919. Now, even an elementary study of geography forces one to consider the whole earth; it forces comparisons between countries, and peoples, and cultures. Geography supplies, therefore, the much-needed foundation for a rounded study of man.

Nevin M. Fenneman, 1919, p. 7. The geographer is, or should be, the greatest insect that carries pollen from field to field.

Albert Perry Brigham, 1922, p.12. It is no disgrace to geography that its progress, as is true of all movements, has been by zigzags. We have wandered more or less from the air line that leads to the goal, but we have been keeping the general direction. There was in the first years of our period strong, and as some thought undue emphasis on physical laws and land forms. But this is the very impulse that not only projected physical features in something of their vast significance, but actually led us into the field of the casual and the human which we now properly regard as of vital meaning. There was no squandered energy.

Nevin M. Fenneman, 1922, p. 23. If America is to contribute anything distinctive to the science of geography it will probably be along the line of closer relationship between human life and its geographic conditions.

Harlan H. Barrows, 1923, p. 3. Geography is the science of human ecology. The implications of the term "human ecology" make evident at once what I believe will be in future the objective of geographic inquiry. Geography will aim to make clear the relationships existing between natural environments and the distribution and activities of man. Geographers will, I think, be wise to view this problem in general from the standpoint of man's adjustment to environment, rather than from that of environmental influence. The former approach is more likely to result in the recognition and proper valuation of all the factors involved, and especially to minimize the danger of assigning to the environmental factors a determinative influence which they do not exert.

Harlan H. Barrows, ibid. The geographic field has shifted steadily from the extreme physical side toward the human side, until geographers in increasing numbers define their subject as dealing solely with the mutual relations between man and his natural environment. By "natural environment" they of course mean the combined physical and biological environments. Thus defined, geography is the science of *human ecology*.

Harlan H. Barrows, ibid., p. 6. Geography is concerned with *place* relations; ecology may well be its organizing concept.

Harlan H. Barrows, ibid., p. 7. If geography be regarded as human ecology, three major systematic divisions of the subject are at once to be recognized, namely economic geography, political geography, and social geography, corresponding to the three great types of human activity that are related to the earth.

Harlan H. Barrows, ibid., p. 8. We come now to regional geography, properly recognized as the culminating branch of the science because it involves facts and principles from all the divisions and subdivisions of systematic geography. I believe that regional geography, even in its widest sense, properly is concerned only with the mutual relations between men and the natural environments of the regions or areas in which they live.

Harlan H. Barrows, ibid. Regional geography comes into its own when the synthetic element is introduced and the different items are studied in the light of their interactions upon one another. Proponents of this view hold, moreover, that such a study of interrelationships in an uninhabited area would still be of the essence of geography. It does not seem to me, however, that geography has any function to perform in connection with such studies of non-human relationships.

Harlan H. Barrows, ibid., p. 9. The center of geography is the study of human ecology in specific areas.

Harlan H. Barrows, ibid., p. 13. I believe that a motivating theme, an organizing concept, is required which shall permeate geography, and give to all its divisions a distinctive point of view. I believe that the problem of the causation of the distribution of surface phenomena, urged by some, and the task of explanatory description of regions, advocated by others, alike fail to meet this requirement, and that the problem of human ecology may have the vitalizing, unifying influence needed.

E. E. Lackey, 1924, p. 67. Concerning the first elemental notion in geography, human activities, we are well equipped with adequately organized facts. Concerning the second elemental notion, namely, earth conditions, we are likewise nicely equipped with well organized data. With this equipment concerning these two elemental factors in geography at our command, we are prepared to consider the third and crowning element, namely, geographic principles. Geographic principles are those relationships among geographic facts that have a universal application. Geography, then, is a study of those universal relationships that exist between human activities and earth conditions. These definitions are ideals toward which to strive. Even now we may consider that the facts of geography, namely those concerning human activities and earth conditions, are exact and organized, but we cannot say that the principles of geography are exact and organized.

Carl O. Sauer, 1924, p. 18. A well-founded objection to much geographic study is based on the fact that the geographer has taken the affirmative side in a debate on environment and has therefore failed to maintain the objective quality of the scientist.

Carl O. Sauer, 1925, p. 348. Geography under the banner of environmentalism represents a dogma, the assertion of a faith that brings rest to a spirit vexed by the riddle of the universe.

Douglas C. Ridgley, 1925, p. 66. A *geographic principle* may be defined as *a fundamental truth concerning the relationships existing between man's activities and his natural environment.* A *geographic factor* may be defined as *an element of the natural environment which influences the activities of man.*

Alton R. Hodgkins, 1925, p. 226. To think geographically is to examine the various situations in which one finds himself, with the purpose of discovering the natural environment factors that give rise to or affect that situation.

G. T. Renner, Jr., 1926, p. 337. ("Commercial Geography.") American geographers have in the past pursued the English and German objectives, but there has gradually arisen in America the idea that geography is the study of the adjustment of man's activities to the environment. The idea of regional study has been incorporated as a powerful aid in geographic organization, while distribution and location are effectively employed in presentation, but neither the regional organization of facts nor their detailed distribution are any longer esteemed as worth-while ends for geographic study. Instead the goal of geographic investigation has become the search for the relationships existing between human activities and the environment, which must be known in order that geographic adjustments may be described and explained.

G. T. Renner, Jr., 1931, p. 33. In its scientific aspect, geography has been defined, and probably will continue to be defined as human ecology. Since the term ecology implies the relations between organism and habitat, it follows that the introductory course in geography must provide training in the discovery and explanation of fundamental ecological relationships. Ecological concepts may be regarded as the epitome of scientific thinking.

G. T. Renner, Jr., ibid., p. 34. The subject matter of geography falls into two fields, one of which deals with the facts of man and his activities and the other with the facts of the natural environment. The purpose of the introductory course is to so handle these two fields that the facts of the former will, in proper measure, be explained in terms of the latter.

A. E. Parkins, 1931, p. 1. Most American geographers hold to the postulate that the behavior of man and groups of men can not be understood and interpreted apart from their environmental setting. We do not accept natural environmental determinism in this postulate but regard the various phases of behavior as adjustments to the environment that man has found works best in his attempts to procure the necessities for the preservation of life in his surroundings.

Alexander McAdie, 1931, p. 91. Geography is no longer taught as a two-dimensional subject, that is, the mere boundaries of certain surface areas. Our school geographies fifty years ago began to describe customs and manners and later chief industries, exports, etc., and so brought in the human factor. We may define geography today as "all that pertains to human environment."

Ellsworth Huntington, 1932, p. 62. Geography has made steady progress from the purely physical to the human aspects. Human geography has progressed steadily from the racial, economic, and political phases to the social, cultural, and intellectual phases. Thus there arises an urgent need for a sound undertaking of the relation between intellectual activity and the environmental background.

W. M. Davis, 1932, p. 228. But I cannot resist the feeling that the forms of the land on which all these human elements are superposed ought to be fairly well understood before the superposed human elements are taken up.

Stanley Dodge, 1932, p. 168. Man as a geographical agent works with his fellows. . . . The solitary man is not only actually a very rare individual, but also geographically a nonentity. It is only when he is integrated with the social and economic fabric of this area, indeed the usual case, that man becomes important. Thus, the geographical expression of that integration . . . is the important thing, for it unites the various parts of the man-made landscape.

William J. Berry, 1933, p. 238. It is substantially my opinion that geography is a natural science. However the matter of the subject is defined—the description of the surface of the earth—it always refers to the surface of the earth and generally to man as well. Now, if it always has reference to the surface of the earth, it is more nearly related to the other earth sciences, e.g., geology, than it is to the social sciences. If it relates to man, it may be either. What, then, is the geographical point of view towards man? Do not geographers consider him as *on* and *of* the earth? I say, yes. If they do not then they are social scientists perhaps, but I deny that they are geographers. The distinction is a subtle one, but I think it is valid.

William J. Berry, ibid. In my opinion geography is primarily a natural science, the science of the earth in its relation to man.

William J. Berry, ibid. I feel that geography definitely belongs with the social sciences but that, of course, many of its data are based upon the findings of the natural sciences. That is true not alone of geography but also of such things as economics, sociology, and the like. All of these require natural science data but express their findings in terms of social reactions. . . .

My chief reasons for feeling so sure that geography does not belong with the natural sciences is that I am afraid if it is put there all that we desired in human emphasis is likely to disappear. We certainly do not want to return to the physiography of a former day.

Preston E. James, 1934, pp. 81–82. There seems at times to be a considerable break between the geographers of an older generation in America, who described their objectives as the study of the influence of environment on man, and the younger geographers who have largely abandoned this description of their field. Some of the younger geographers have adopted the broader traditional definition of their objectives as the description and interpretation of the face of the earth.

Wellington D. Jones, 1934, p. 94. All commonly accepted definitions of geography include the study of "human occupance of regions," and that no other recognized discipline is so particularly concerned as is geography with the intimate association of peoples with the areas they occupy.

Willis H. Miller, 1936, p. 279. Modern geography admittedly is concerned with the study of ephemeral relationships existing between human activities and the natural environment.

Willis H. Miller, ibid. Current events vary from the Dionne quintuplets to the Einstein Theory, both of which are equally unintelligible to the average reader. As geographers we fortunately are able to restrict ourselves to those comparatively few current events having a high degree of that priceless ingredient, geographic quality. To have well-marked geographic quality a news item must show a close and fairly direct relationship between human activity and the natural environment.

Fred B. Kniffen, 1937, p. 163. I believe that the primary concern of cultural geography is with the nature, genesis, and distribution of the observable phenomena of the landscape directly or indirectly ascribable to man, and of course including man himself.

John Leighly, 1937, pp. 134–135. Is it too great a strain on long-established and familiar words to call the study of localization of the cultural immobilia with which the geographer is likely to work "topography of art"? No violence is done either to language or to the tradition of learning if the expression "topography of art" is used in the sense of scientific discourse concerning the localization of artifacts, however far the attempt to explain this localization may be carried. The topography of art therefore includes most of what is commonly subsumed under such a designation as cultural geography.

Wallace W. Atwood, 1940, p. 44. ("Fundamental Basis.") No one will ever go very far in the study of geography who does not understand the fundamentals of mathematical geography, the origin of the present land forms, the changes in climate, the nature and distribution of the natural resources in the ground, the differences in soils, in the major plant societies, and the nature and tendencies of the human beings who are at work in the various habitats on this planet.

John Leighly, 1940, p. 64. The doctrine of environmentalism represents an effort to carry over to cultural phenomena the principle of causality developed and successfully applied to phenomena of the physical world. Because history is the scholarly field that until recent times concerned itself primarily with social phenomena, the environmentalistic doctrine has generally been directed toward the events that history has recorded. This doctrine has therefore appeared mainly as a theory of history, competing with other theories of historical causation. Apparently it has been particularly prominent at junctures in the history of western thought at moments when the principle of physical causation has enjoyed high esteem.

John Leighly, ibid., p. 65. As historians have receded from a position that required them to provide a mechanism for "explaining" their accounts of the past, they have ceased to feel a need for any "theory of history." Sociology, as a self-conscious discipline, has also receded from the position Herbert Spencer gave it, that required it to be a necessarily rational physics of society. At the end of the nineteenth century geography was left holding the environmentalistic bag while history, ethnology, and sociology went about other business. In the United States Miss Semple gave the environmentalistic dogma in geography a much more drastic formulation than her teacher Ratzel had given it. Her essentially poetic utterances were obsolete as science when they were announced. They were built on ethnologic theory that was current in the first half of the nineteenth century.

Robert Peattie, 1940, p. 68. The fields of cultural geography and environmentalism are complementary. The factors which are causal in civilization are heredity, environment and human choice. Let these three variables be the dimensions of a cube and all conditions of a culture can be plotted into the cube according to the strength of the three variables. The concept of determinism is possible only when two variables are reduced to zero. The fact would then be plotted in an extreme corner of the cube.

Rafael Picó, 1941, p. 300. As a cultural subject with practical applications, geography has today earned a deserving place in American higher education. As a science stressing the areal expression and correlation of human, economic, and physical phenomena, it offers a distinct point of view and a special technique. Thru its research, geography has added a wealth of data and interpretation. However, perhaps its greatest contribution to an era of extreme specialization is its ability to provide a much needed link between the modern physical and social sciences.

Ellsworth Huntington, 1942, p. 1. Some even said that unless man is specifically included in a given problem it is not geography. Many thought that the description of a factory was geography. In correcting this extreme view we need not become as geological as Professor Davis, but we must heed that great geographer's insistence on the reciprocal nature of physiography and ontography. It is regrettable that geographers do not use these two terms more commonly, for we need more emphasis upon the dual nature of our subject.

Derwent Whittlesey, 1943, p. 6. ("Place of Geography.") Only geography offers a base fixed in nature from which the historical, political, economic, and sociological effects of . . . change can be measured.

Isaiah Bowman, 1945, p. 215. An imaginative grasp of space and time is required of the modern scientific geographer. No other science puts humans and earth in their regional frameworks and tries to appraise the systems of resource-use that men have created.

Isaiah Bowman, ibid., pp. 214–215. Using tools of physics, chemistry and biology, geography puts together the two parts of a region that have human meaning. All of the great geographers of history discovered human meanings, not merely coasts and mountains and river systems.

Richard Joel Russell, 1945, p. 306. Herodotus followed the core of geography, as did Strabo several centuries later. Each set out to describe the earth and its inhabitants and to draw significant conclusions from their examinations. Men have always been interested in that subject and always will be. The domain of the geographer has expanded vastly since the time of the Ancients, but the central theme of the subject has remained the same. Paths divergent from the core have led to many fields now established permanently under other names. The whole growth and subdivision of knowledge into fields today is so complex that most paths diverging from the central theme of any ancient branch of learning lead into disciplines preëmpted by others.

Richard Joel Russell, ibid., p. 306. The interest of the geographer in peoples arises primarily because they occupy and profoundly affect the surface of most parts of the earth. Yet the presence of man is not a necessity in defining the central theme of geography. Few will deny that there is a geography of Antarctica because that continent is uninhabited. The boundary of geography is not reached where the environment ceases to act on man, or man on the landscape.

Eugene Van Cleef, 1947, p. 92. I do not hold geography to be a social science. It is an earth science of tremendous significance to a correct interpretation of life's activities—plant and lower animal as well as human. We do not style an entomologist a social scientist when he delves into the mysteries of the social organization of ants, nor do we classify as a social scientist, the zoologist who concentrates upon genetics. Need a geographer who inquires into the relationship between the animal man and the physical earth be thrown into the same category with the economist or sociologist, that is, into the realm of the social scientists? As a matter of fact, the expression social science is a paradox, for the society of man has yet to be shown as a phenomenon whose behavior can be predicted with finality.

Robert S. Platt, 1948, p. 351. ("Environmentalism Versus Geography.") The extreme claim of environmentalism—that the natural environment controls or determines human life—is recognized as false.

Robert S. Platt, ibid., p. 353. The types of natural environment in physical science are actually expressions of social science in their dependence upon cultural criteria. We may describe a region of rolling plains or of dry subtropical climate in physical terms of altitude and annual rainfall; but our distinctions are made through human experience and for human occupance under known forms of culture.

Richard Joel Russell, 1949, p. 11. To me, geography is essentially the study of places, and of people because they have so much to do with modifying places. From this core I visualize our field as radiating in all directions, and for various distances, toward the cores of other disciplines. Questions of boundaries always seem pedantic in comparison with questions concerning the cores of disciplines.

J. Wreford Watson, 1953, p. 315. Vidal de la Blache of France, Mackinder of Britain and Huntington of the United States, [*sic*] saw geography as a meeting place, rather, of history, economics, politics, anthropology or sociology. They stressed human or social geography, as distinct from physical geography, and while agreeing that geography must have a firm foundation in the physical sciences, believed that its real application was in the field of the social sciences.

Arthur N. Strahler, 1954, p. 5. The empirical-quantitative system is indispensable in the actual practice of geography where the influences of the physical environment upon man, his economy, history, and institutions are being studied.

Fred Kniffen, 1954, p. 222. Cultural geography is deemed a social science. As such it should first order and interpret its data, man's works appearing on the earth's surface, to contribute ultimately to the understanding of human behavior even to the point of prediction. Man-land relations must be the aspect of human behavior to be studied by geographers, and prediction falls in the realm of what man is going to do with respect to earth qualities.

Edward T. Price, 1955, p. 64. Geographers in the United States seem to have committed their profession to a shepherding of natural resources. It is fitting that students of the earth's surface should keep tab on its riches, take part in planning its management, and apprise the coming generation of its potentialities.

Emrys Jones, 1956, p. 369. For some years past it has been assumed that geographical determinism is a discredited mode of thinking. If a geographer believed in such a heresy he was careful to maintain his respectability by hiding the fact. The majority who could not in good faith dismiss it in its entirety, took refuge in "possibilism." At first glance possibilism seemed to give that scope for freedom in man's actions which had been denied in a strict environmentalism; but although it spoke in terms of collaboration between man and nature, and even stressed that man exercised a choice and thereby brought in freedom of will, it was clear that the choice was one within limits, and that those were set by nature. Although man was not "fatally determined" he was, it seems, "circumscribed," and his freedom was more apparent than real. The famous words, "There are not necessities, but everywhere possibilities," lose something of their all-embracing freedom when we remember that to Febvre different regions offered possibilities differing sufficiently in number and quality to warrant a hierarchy. . . . Man chooses, but only from the range with which nature presents him. This, after all, is not very dissimilar from the modified determinism of Griffith Taylor. In the latter there is, it is true, a Master Plan, and on it a path which is indicated by nature and which man would do well to learn. But even if, as possibilists suggest, there are several paths—i.e., several possibilities—from which man can choose one, does not this also suggest a plan?

Emrys Jones, ibid. Indeed . . . possibilism proves to be more a variant of determinism than of libertarianism: but so innocuous a variant did it seem that geographers on the whole were content, happy in the thought that determinism as a doctrine belonged to the past.

J. Sonnenfield, 1960, p. 172. The abstract idea of environment has little meaning for the human geographer who tends to view environment primarily in its relation to man. The concept of an environmental potential can never be absolute since it varies according to man's technology and "needs." As these change so does his concept of the utility of the environment change.

Brian J. L. Berry, 1964, p. 3. Geography's integrating concepts and processes concern the worldwide ecosystem of which man is the dominant part.

William D. Pattison, 1964, p. 214. One can only regret that this narrowed version of the man-land tradition, combining all too easily with social Darwinism of the late 19th century, practically overpowered American professional geography in the first generation of its history. The premises of this version governed scores of studies by American geographers in interpreting the rise and fall of nations, the strategy of battles and the construction of public improvements. Eventually this special bias, known as environmentalism, came to be confused with the whole of the man-land tradition in the minds of many people.

Yi-Fu Tuan, 1965, p. 652. From a humanistic standpoint the expression "man and land" is an invitation to understand man's attitude toward his environment.

F. Hung, 1966, p. 343. Quantitative analyses of the "man-environment system" and of the subsystems of the physical environment in place of largely descriptive geography have hardly started on any large scale.

Ian M. Matley, 1966, p. 111. Perhaps the most important lesson of all arising from the debate on the role of the geographical environment in the development of society is that the concept still exists as a central theme of Soviet geography at a time when many American geographers can feel, and state, that geography, to the satisfaction of much of the profession, has finally rid itself of what is thought to be the last traces of geographical determinism and has exorcised its ghost. There are, of course, American geographers who share preferences for some degree of environmentalism or possibilism, but they now are less outspoken. Indeed, so thoroughly has this exorcism taken place that many geographers experience a strong feeling of discomfort when any suggestion is made that the natural environment may play any role other than a purely passive one; any mention at all of the natural environment is embarrassing to some. Perhaps some of us have leaned too far backward to avoid the stigma of environmentalism.

Gordon R. Lewthwaite, 1966, p. 1. ''Environmentalism'' and ''determinism'' are terms covering varied concepts. Environmentalism included both environmental determinism and the environmentalist definition of geography as the study of man-environment relationships. These were not identical: most who accepted the environmentalist definition preferred possibilism to geographical determinism. These two positions were not consistent extensions on the metaphysical concepts of free will and determinism: possibilism denied environmental control but not necessarily other determinants, whereas geographic determinism conflicted both with possibilism and all other particular single-factor determinisms. But since environmentalists never completely excluded cultural factors, they differed from possibilists and especially probabilists only in degree.

James D. Clarkson, 1970, p. 700. Placing an ecological approach in the general framework of American geographic thought indicates the usefulness of distinguishing two trends in the development of this thought—the one ecological, the other spatial. American geography tended to reject the ecological approach because it was identified at an early period with environmental determinism. A spatial, non-functional, approach became dominant. Although the two approaches are two ends of a continuum, and thus connected, they arise from and lead to different sets of questions which involve different approaches and different bodies of theory. The ecological approach may be divided into four imprecise types—biological, human, cultural, and urban-political. The cultural-ecological approach is particularly useful in analyzing obstacles to innovation acceptance in agricultural development because it emphasizes the analysis of existent systems from different viewpoints. Four sets of reality, or viewpoints, can be distinguished in this context—that of the ideal-set of the cultivator. Only when the overlaps and conflicts of these sets are recognized can a realistic appraisal be made. This is only a single instance of the potential of an ecological approach. Spatial theory and ecological theory have not yet been joined. The evident usefulness of both indicates the importance of attempting such a joining, and the futility of arguing for the ascendance of one over the other.

C. O. Sauer, 1971, p. 254. As we approach the end of the nation's second century we are in grave and unresolved crisis of destructive exploitation and urban malaise. Instead of moderation of material satisfactions advance of technology has accustomed the people to want more things of brief appeal. The automobile industry became the giant of the nation by planned obsolescence. We have built an economy based on waste and boredom and both are overtaking us.

Norton Ginsburg, 1973, p. 4. The low priority placed upon intensive research on non-Western societies is almost a scandal. The relative neglect of ecosystemic approaches to social problems at a time of deepening ecological concern and even crises has helped set the profession back years. The emphasis on geometrics as explanation rather than description is a spectre that will haunt our efforts increasingly as time goes on.

Norton Ginsburg, ibid., p. 20. The fact is that there is no substitute for understanding how people see the world around them and what sorts of priorities they are likely to develop. If these are understood, then it may be possible to describe and understand the kinds of environments they will create for themselves. If the environmental systems within which men live and the forces that impinge on them are understood, then one is more likely to forecast the future forms of those environments, or at least, through simulation, to estimate what the more likely consequences of change will be.

Edward J. Taaffe, 1974, p. 3. The man-land view had many strengths. Under the strict definition it could be clearly stated and understood, and it did identify a set of problems. This clarity of problem identification enabled the geographers of the twenties, although a few in number, to have remarkable impact on public policy.

Edward J. Taaffe, ibid., p. 5. The man-land view also had several negative aspects, however. For one thing, it begged the investigative question and forced the investigator to look for relations between man's activities and some aspect of the physical environment.

Edward J. Taaffe, ibid. Another weakness of the ecological view as it existed in the twenties was its failure to lead to cumulative generalizations. What one geographer found out about the effect of environmental features was seldom referred to in the next geographer's study.

Marvin W. Mikesell, 1974, p. 2. The current quest for environmental understanding promises to be different, for there is ample evidence of a more active and creative response to the several issues raised during the past few years and perhaps even an aura of confidence about the advantages of geographic education and the power of geographic tools.

Marvin W. Mikesell, ibid., p. 4. For most geographers the alternative to environmentalism was to disregard the ecological commitments of the discipline in favor of distributional or locational studies that paid only perfunctory attention to environment. Thus although land use and land forms might both be mapped, the only generalization offered about a connection between the two would be one of coincidence or correlation, e.g., steep slopes have different uses than gentle slopes. Not a few geographers simply ignored environment and devoted themselves to functional or classificatory studies of economic activities. The excitement created by the concept of "nodal regions" and the vast energy expended on the mapping of land use were sufficient to discourage more than a casual second look at environmentalism.

Marvin W. Mikesell, ibid., p. 7. The most serious failing of the early environmentalists was their tendency to regard nature and culture as separate entities or opposing forces rather than as interlocking components of such systems. The trials and errors of their effort have taught us that man acts in and upon nature rather than merely in response to the dictates of nature.

James Parsons, 1977, p. 1. Geography, so magnificently interdisciplinary, seems an ideal vehicle for the joining of hands of science and humanism, including the taking of moral positions on environmental and spatial issues.

Melvin G. Marcus, 1979, p. 530. The pressing problems of energy, environment, and urbanization provide a timely laboratory in which human and physical geographers can perfect the human-environment paradigm.

Karl W. Butzer, 1980, p. 273. There is then every reason to believe that a continuation of present environmental trends will spell an economic crisis of global proportions within the lifetimes of our children and grandchildren.

John Fraser Hart, 1982, p. 6. The usefulness of "landscape" as a technical term was diminished somewhat by the fact that no one ever really knew for sure what a landscape is, or is supposed to be, and no geographer has ever been able to define the term to the satisfaction of anyone other than himself.

John Fraser Hart, ibid. Even environmental determinism, which aroused such fierce emotions among geographers a generation ago, and has been almost completely forgotten today, began as an attempt to make geography more scientific by replacing teleological explanations with the explanations of natural science. It stemmed from Darwinian notions of natural selection and the survival of the fittest in a given natural environment. The determinists saw the natural environment as cause and human activities as effect; nature was the physical mold in which human activities were formed and shaped.

John Fraser Hart, ibid. The determinists developed a useful standard checklist that included geology, landforms, climate, soils, and vegetation as major elements of the natural environment that were assumed to determine human behavior. They considered natural features so important that they were worthy of study in their own right, but they treated human activities as an afterthought, a mere nuisance, and some geographers even went so far as to propose that human activities should be completely eliminated from geography on scientific grounds.

John Fraser Hart, ibid., p. 24. Geographers, like the mythical giant Anteus, derive their strength from contact with the earth. Anteus became stronger each time he was hurled to the ground, and Hercules finally was able to vanquish him only by holding him aloft. Perhaps we should designate Anteus as one of the patron saints of geography, because we too grow weaker the farther we depart from Mother Earth.

William M. Denevan, 1983, p. 399. The point of view, "paradigm" if you will, of traditional cultural geography in the U.S. seems to be one in which human behavior vis à vis the physical world is explained in terms of learned group experience; i.e., culture.

Richard Morrill, 1983, p. 5. . . . geography can make a distinctive contribution in its capacity to describe and analyze how diverse physical and human processes do interact to produce particular regional landscapes, cultures and places.

Shlomo Hasson, 1984, p. 11. In its challenge of rational positivism, humanistic geography questions the very possibility that people's relationships with place and space can be understood without grasping their feelings, emotions, values and intentions. Thus, in contrast to rational positivism which is geared towards "explanation in geography," i.e., the derivation of laws concerning the occurrence of given phenomena in space, humanistic geography displays an intellectual effort which aims at understanding geography, i.e., grasping the inner meaning of the relationship between people's experience and environment.

Gilbert F. White, 1985, p. 10. Fifty years and one month ago I came to Washington to help save the nation. The sojourn was to be for six weeks working with Harlan Barrows before returning to the university to write a doctor's dissertation. The six weeks stretched gradually to eight years. Like many others, young and old, I came with a modest but fervent sense of mission. It was the New Deal. We were there to get the country going again, to right the inequities of the Great Depression, and, on a more earthy level, to remedy the onslaughts of erosion, drought, and flood. The lights burned in offices along Pennsylvania Avenue at night, including weekends. There was enthusiasm, willingness to experiment, and a belief that somehow ways could be found to put the nation back on track.

# PART THREE

# Area Studies Tradition

## REGIONAL VERSUS TOPICAL GEOGRAPHY

Charles Redway Dryer, 1912, p. 73. As to method of treatment and presentation, geography has two distinct and contrasted phases, general geography and regional geography. General geography discusses features and processes, seasons, mountains, rivers, ocean currents, coal, iron, deserts, forests, corn, cotton, agriculture, manufacture, trade, education, politics, and so on through the whole geographic category, as each occurs over the whole world. Regional geography discusses the same features as they occur in a definite and limited region, as the British Isles, the United States, New England, the glacial drift plain. In most textbooks of physical geography the treatment is general; in textbooks of human geography it is prevailingly regional.

Eugene Van Cleef, 1915, p. 241. If the object of this paper were merely a recapitulation of what has been written upon the subject of the sugar beet or beet sugar, it would hardly be worth while. The literature is very extensive in quantity, but shows no very great variety. In essentially no instance is the subject treated from a general geographic point of view.

Alfred Hettner, 1927, p. 1308. There is no proper distinction between regional and general geography—the ultimate objective and fundamental concept of geography lies in the region.

V. C. Finch, 1930, p. 182. I would without hesitation choose a regional rather than a topical organization of material as the more broadly cultural and the better suited to the education of the beginner. A regional treatment better presents to the casual student the peoples and problems of other lands. Here may be painted the picture of the life of people in its setting of the characteristic elements of site and situation. This appeals to the beginning student and it lays the foundation for the addition of facts gleaned later from general reading and obtained thru personal experience in travel. Also this presentation is better adapted to the needs of the student who proposes to specialize in the field of geography. It introduces him at once to the regional way of thinking which is essentially geographical.

It gives him the elements of a method which may later be developed into his own regional method.

William Morris Davis, 1932, p. 217. Let no young geographer, therefore, be disheartened when he discovers the multiple nature of his subject or the impossibility of mastering all its content. Let him first secure a good understanding of the fundamental components of Geography, and then select some special component to study systematically in its occurrence all over the world, and some special area of the world in which to study all the components in their regional associations. He will thus, on reaching middle age, have gained expert acquaintance with his selected area; and that will suffice to make him through his later years a real geographer.

W. Elmer Ekblaw, 1937, p. 214. (''Attributes of Place.'') If place be accepted as the essential concept of geography then the content of geography can be logically and reasonably determined. Its discipline can be definitely established upon the criterion of place. Its field can be definitely restricted to the attributes of place, and by the boundaries set up in that criterion. Facts or principles that involve the concept of place in any of its attributes or relationships, thus become definitely geographic, and amenable to geographic discipline. Fact or knowledge that does not apply to an attribute or relationship of place is not properly geographic, however significant it may be in itself.

W. Elmer Ekblaw, ibid., p. 219. The topical approach, particularly in the several aspects of human geography, offers the gravest hazard of departure from geographic discipline, in that the attributes of place are more or less concealed under a mantle of social, economic, and political phenomena only remotely, indirectly, or indistinctly related to place.

V. C. Finch, 1939, p. 4. Is regional geography as a field of learning amenable to the methods of the sciences, or must it, as some are maintaining, turn mystical and produce fruit only in the realm of literary art?

V. C. Finch, ibid., p. 23. Regional geographers do not maintain that *all* workers under geographical discipline should perform as regionalists or even that they themselves should always treat of the regional entity in its totality.

Richard Hartshorne, 1939, p. 260. The ''classical period'' in the development of geography may conveniently be considered as terminating with the death of both Humboldt and Ritter in 1859. During the following decade academic geography was dominated by the school which Ritter had founded. Since his followers tended to emphasize the ''historical'' aspects of the field even more than had Ritter, one may speak of a drift away from systematic geography to a regional geography primarily concerned with man.

Richard Hartshorne, ibid., p. 550. The failure to understand that geography is to be defined essentially as a point of view, a method of study—just as all science is a method of study—has caused many to suppose that the growth of the daughter sciences had left nothing for the parent science to do. Attempts have been made to save the day by claiming for geography a particular type of phenomena, such as relationships between man and nature, or by searching for new objects of study which no one else has previously considered worth studying, or by attempting to metamorphose abstract concepts of area into concrete objects. Each of these efforts, to a greater or less extent, has caused geography to depart temporarily from its path of development in directions which have proved, or will prove, to lead either into fields that other sciences will not cede to geography, or into the bog of mystical thinking. Geography is a study which looks at all of reality found within the earth surface from a particular point of view, namely that of areal differentiation. This might be called the *position* of geography as a field of knowledge. More significant to the general question is the *character* of geography as a field in which knowledge is acquired.

Richard Hartshorne, ibid., pp. 590–591. The divisions of systematic geography . . . correspond to the divisions of the systematic sciences and there is inevitably close relationship between each branch of systematic geography and the corresponding systematic science. This relation is not accurately expressed by the phrase "neighboring sciences," since geography is not a branch of science situated beside the systematic sciences, but represents a point of view in science which cuts through all the systematic sciences. There is, therefore, no line separating systematic geography from the systematic sciences, but there is an essential difference in the point of view which must be maintained by the individual geographer who wishes to do geographic work rather than work in some other branch of science.

Richard Hartshorne, ibid., p. 636. Geography takes from the systematic sciences all knowledge that it can effectively utilize in making its descriptions of phenomena and interpretations of their interrelations as accurate and certain as possible.

Richard Hartshorne, ibid., p. 641. In regional geography all the knowledge of the interrelations of all features at given places—obtained in part from the different systems of systematic geography—is integrated, in terms of the interrelations which those features have to each other, to provide the total geography of those places.

Richard Hartshorne, ibid., p. 642. The greatest differences in character within geography are found between the two major methods of organizing geographic knowledge—systematic geography and regional geography—each of which includes its appropriate part of all the special fields. In addition to the difference in form of organization in the two parts, there is a radical difference in the extent to which knowledge may be expressed in universals, whether generic concepts or principles of relationships.

Richard Hartshorne, ibid., p. 644. Systematic geography is essential to an understanding of the areal differences in each kind of phenomena and the principles governing their relations to each other. This alone, however, cannot provide a comprehension of the individual earth units, but rather divests them of the fullness of their color and life. To comprehend the full character of each area in comparison with others, we must examine the totality of related features as that is found in different units of area—i.e., regional geography.

Wallace W. Atwood, 1940, pp. 44–45. ("Fundamental Basis.") We may think of the goal towards which we are moving as an understanding of the great human dramas which are in progress in the various geographic regions. Those dramas are continuous, and all who are present are actors.

The superstructure in our science is human geography, but unless the superstructure rests upon a well constructed foundation it will crumble, and any who venture to build such a superstructure without a thorough knowledge of the fundamental phases of the science are doomed to disappointment. We should have in our professional group specialists in each phase of geography. We should welcome research work carrying forward the frontiers of knowledge in any one of those phases. They are essential parts of geography.

W. William-Olsson, 1940, p. 420. A city may be studied geographically from two different points of view: the city as such or in relation to its surroundings.

Edward A. Ackerman, 1945, p. 135. We have specialists on Latin America, Africa, Australia, the South, or the Mediterranean, who do not have the advantage of the experience and data which should, and can only be provided for them by students of industrial geography, political geography, settlement, or a dozen other subjects—data without which they can never hope for full understanding of a region or any part thereof. In general we are still relying on a country doctor method in the face of increasing intricate problems.

This past method may be one reason for the concern which some critical students display for the development of regional science. Finch, for example, says that "perhaps . . . the regionalist may be forced to admit that the complexities of man and nature in area are too much for his abilities at rationalization." Hartshorne, in turn, feels that " . . . no matter how many regions are studied, no matter by what methods, no scientific laws will be forthcoming." These observations may be correct, and certainly they are supported by experience thus far. Are we not too near the beginning of our road, however, to make pessimistic forecasts of any sort? Should we not recognize that we merely are at the end of a stage, and probably poised to attack from another angle of being so ready to admit our limitations?

Edward A. Ackerman, ibid., p. 138. The possibilities inherent in the cooperative systematic attack on regional problems have been obscured in the past not only by the general academic individualism, but by the repeated statement that a group of systematic geographical studies taken together cannot be added to make regional geography. The individual regional geographer, it is maintained, must be present to dispense the nuptial interpretation. Theoretically, however, if the systematic studies are *perfect*... they could be added together to form regional geography. The fact that such an addition has not been achieved in the past can be explained by the near or total absence of cooperative research among systematic geographers. It is unlikely that an individual specialist can arrive at perfect understanding of all the correlative relationships connected with its subject without close contact with other students in neighboring specialties. But it is even more unlikely that an individual specializing in all relationships (the regional geographer of the immediate past) will attain that perfect understanding.

Edward A. Ackerman, ibid., p. 140. Insistence upon the duality of geography, and the attraction of holistic regional study for planners of curricula and research has left our profession in a situation where we must contemplate a radical revision of approach. For those of us who accept the value of wartime experience, the handwriting on the wall is very clear. Geographers for the most part will struggle forward with primitive instruments if they do not redouble efforts toward systematic research, formulation of systematic principles, and provision of adequate training in systematic geographic methods. Believers in the validity of regional geography have as much at stake as anyone else, perhaps more, for the ultimate in regional geography quite clearly must wait upon progress as yet unattained in systematic study.

Edward A. Ackerman, ibid., p. 141. We might also say that human geography is in an unenviable situation, so long as there is the present emphasis upon regional methods of instruction. Many of our graduates know a little about a lot. They and their studies easily give a justified impression of superficiality to outsiders, a confused illustration of our ultimate objectives, and an example which encourages all sorts of incompetents to try their hand at geographic interpretation. Human geography (an ultimately regional geography) will never be accepted as a mature scholarly discipline until a more thorough systematic literature begins to take shape in it. That in turn is at least dependent on a reshaping of curricula in graduate schools.

Edward A. Ackerman, ibid., p. 143. We are at a stage in the development of geography where systematic methods should be emphasized in training and research.

G. Donald Hudson, 1951, p. 114. Geography is concerned with the totality of area. To be studied, the totality of area must be broken down into its components, each becoming the center of attention. Thus in its approach geography is systematic. Studies that confine themselves to this approach we label *systematic*. When the findings and understandings with respect to these components are integrated geographically—that is, areally—they form the base from which areal findings and understandings are derived and formulated. Studies that perform this function we label *regional*.

Preston E. James, 1952, p. 198. The topical specialist in geography can be distinguished from scholars in neighboring disciplines because of the focus of his interest on the differences developed from place to place on the earth by a process rather than on the process itself.

Fred K. Schaefer, 1953, p. 249. What may one infer from all this for the future of geography? It seems to me that as long as geographers cultivate its systematic aspects, geography's prospects as a discipline of its own are good indeed. I am not so optimistic in case geography should reject the search for laws, exalt its regional aspects for its own sake and thus limit itself more and more to mere description.

J. Russell Whitaker, 1954, p. 237. We need a history of geographic thought that shows the time setting of each major essay in geographic philosophy and methodology. What a man writes from his heart today is not what he would have said twenty years ago or what he would feel compelled to say ten or twenty years hence. We utterly misjudge the view of geography as "human ecology," for example, unless we see it in its nearness to "geographic influence" and "content of area." We fail to understand a critique of regional geography published in 1937 unless we see how far the pendulum had swung toward regional description in the years immediately preceding, for the Second World War has since carried us far toward an appreciation of the topical approach.

Carl O. Sauer, 1956, p. 294. If we prune out a lot of the regional work now spread through our curricula we shall also urge ourselves to move the topical courses out of their present obscure corners. Topical courses have the advantage that they are analytic, and their elements may be scrutinized at any scale of inspection and by more or less adequate techniques. In the education of the student and his postgraduate development the topical inquiry is attainable and rewarding.

Carl O. Sauer, ibid. I am becoming more and more doubtful that regional studies are for the beginner in research. The more I see of regional theses, with their descriptions and classifications, and dot maps, that are possibly useful but are mainly secondary collections of presumed facts, the more do I wish that this time and energy had been focussed on some topic which is a problem. What problems are stated and at least partly resolved in an average regional thesis? The incipient regional geographer is either sadly at a loss to determine what he wishes to describe, or else follows a routine grouping of data which depresses his job to pedestrian performance. One end of geographical knowledge is comparative regional understanding; I don't agree in the least that it must be the only end, to which topical studies are only considered as building stones. I'll commit myself further and say that if most younger students stayed on the trail of themes rather than of regions, our contributions to knowledge would be more numerous and of a higher order.

Brian J. L. Berry, 1964, p. 9. If the object of systematic geography is to find those fundamental patterns and associations characterizing a limited range of functionally interrelated variables over a wide range of places, the object of regional geography is to find the essential characteristics of a particular region—its "regional character" based upon the localized associations of variables in place—by examining a wide range of variables over a limited number of places.

John Fraser Hart, 1982, p. 14. The regional side of geography is concerned with patterns, associations, and synthesis; the systematic side is concerned with analysis of the processes that help us to understand and explain these patterns and associations; and they are two sides of the same coin, geography.

John Fraser Hart, ibid., p. 15. Some geographers are more comfortable with systematic geography, and some are more comfortable with regional geography.

John Fraser Hart, ibid., p. 18. We need a unifying theme that ties together all of the wildly disparate phenomena with which we deal, and I know of no other theme that is even remotely so satisfying as the idea of the region. All of the different strands of geography converge when we try to understand a place, an area, or a region, and not one of them can be ignored, which is further evidence of their relevance. We need them all in order to understand regions, and we need the concept of the region in order to understand why we need the diverse and variegated systematic subfields of geography and we hobble along like a one-legged beggar if we try to make do without either.

## REGIONAL DESCRIPTION AND EXPLORATION

Henry G. Bryant, 1913, p. 111. With experienced topographer, two Newfoundland canoemen and the natives . . . , party started in canoes up the river on July 12th. A river flowing in a narrow valley between ridges of Archaean rocks. Uniformity of sky line owing to glacial action. Difficulties of river travel increased, by desertion of Indians and physical disability of one of the canoemen. A ten mile portage. Difficulty of locating Indian route of travel. Source of river reached. Compelled to turn back here owing to continued illness of canoeman, when within ten days of the waters of Hamilton Inlet, the objective point of the journey.

Sumner W. Cushing, 1913, p. 43. For the geographer, a pleasing approach to Japan is through the far-famed Inland Sea. The route runs from the west gate, on the north side of which stands Shimonoseki—the Gibraltar of that oriental Mediterranean—some 250 miles to Kobe, at the east end of the Inland Sea, where his ocean steamer finds it profitable to make the first stop. Throughout this long stretch, if he is fortunate in weather and if the steamer's schedule covers most of this route in daylight, the geographer will receive a lesson of the highest interest, and at the same time be struck by all those more aesthetic appeals of color, form, and life, for which this picturesque inland sea is noted.

John W. Herbert, 1913, p. 241. Stimulated by what he saw on a recent visit to the Panama Canal the writer's desire was aroused for more information than can be derived from a short visit. The following pages are the result of his inquiries. Although making no pretense to scientific treatment they are offered in the hope that they may nevertheless prove of interest.

Leon Dominian, 1913, p. 576. I propose to describe the Balkan Peninsula this evening as a remarkable highway. Before mankind had begun to record its past this region had afforded a natural passage, by virtue of its geographical situation, for the westerly migration of Asiatic races fleeing from the aridity of their land of origin. Throughout historical times it has constituted, with Asia Minor, a natural bridge joining the East with the West. To-day the region bids fair to maintain the same rôle, with the difference that the trend of human flow appears destined to be directed toward the east instead of toward the west.

Osgood Hardy, 1914, p. 500. The object of this report is to give as much information as possible about the portion of south central Peru represented by the Departments of Cuzco and Apurimac, from the standpoint of their commercial and industrial possibilities.

Carl O. Sauer, 1915, p. 343. At the mouths of the tributaries is a zone of grassy swamps. Above, there are luxuriant forests, which are however in many cases merely a screen in front of the sago swamps that intervene between river and upland. On the upper tributaries natives were found who had never heard of white men and were unacquainted with iron.

Godfrey Sykes, 1915, p. 758. The theory sometimes propounded that the early Spanish navigators were influenced in their opinions as to the insularity of California by having penetrated by water into the basin which we now know as the Salton Sink, may be dismissed at once as highly improbable; for even if this depression was at that time filled by the overflow waters of the Colorado River (which is not in itself at all impossible), the fact that it was entered by stemming a swift fresh-water current, and that once reached, the great tides experienced in the gulf were totally absent, would at once show any voyager that he had left the sea and reached a lake.

Preston E. James, 1925, pp. 212–213. After a short voyage across the quiet waters of the Paria Gulf the ship approaches the site of Trinidad's famous Pitch Lake. A long pier, nearly a third of a mile in length, provides the only place in this part of the world where large ocean ships can tie up at a dock. Beyond the pier lies the western shore of Trinidad, low, green, sparkling after the showers from the towering cumulus clouds which are constantly forming on the eastern side of the island and drifting westward over the gulf. Like all tropical islands, Trinidad looks romantic, mysterious, from the deck of the ship. The imagination pictures strange scenes set among the beautiful palms or in the "High Woods." But a closer acquaintance with Trinidad and its people reveals a life very similar to our own: men moved by the same desires, interested in much the same things, influenced by the same trends of thought. To them the tropical setting is common and uninteresting; the regions with long, cold nights, and glistening snowy days appeal to them as the sources of romance and mystery.

Robert S. Platt, 1926, pp. 20–21. Theoretically Central America is a link connecting North and South America: practically it is a row of separate countries each facing the sea highway, each like a house with its own front path down to the highway and its own side fences separating it from its neighbors. There may be much going and coming in every household, but anyone going the length of the block had better go along the highway than try to make his way across back yards and over fences.

Hugh H. Bennett, 1926, pp. 63–64. From the deck of a steamer Central America presents an unfavorable impression as to possibilities for large agricultural enterprises because of the prevalency of rugged topography. Steep hill and mountain land come to the sea in many places, and occasionally lofty volcanic peaks are to be seen rising from the distant cordillera to heights in some instances of nearly 14,000 feet. Nor is the impression removed by closer contact with the terrain, certainly not in many localities. There are numerous areas of smooth country to be sure; flat coastal plains, river deltas, and broad stream bottoms; undulating savannas and high llanos; flattish and gently rolling plateaus and undulating to rolling high valley lands. Nevertheless, one is seldom beyond the sight of steep hills and precipitous mountain sides. In the Canal Zone, for example, where the maximum elevation is only 1,223 feet, and the average height of land less than half this, the country is exceedingly uneven of surface, and the traveller finds at the foot of every hill another hill to ascend, save for an occasional strip of stream bottom and the smoother savannas of the Pacific side.

O. D. von Engeln, 1926, p. 274. What, actually, are human geographers in their brightest dreams striving to achieve? In essence exactly that which the novelists have been endeavoring to do for a hundred years or more.

Mark Jefferson, 1927, p. 31. Genial, lordly whites, by the hands of outnumbering slaves, grew rice on the delta of a mighty river, or cotton of longest fiber on islands or higher parts of a coastal lowland where forests of longleaf pine are interrupted by linear cypress sinks that mark old earthquake fissures along the coast. Proud their mansions, well built by skillful slaves who had been trained by owners careful of their human property. The slaves, though coerced to much toil, maintained a certain happiness with a lowly standard of life.

Ellen Churchill Semple, 1928, p. 420. The Mediterranean climate encouraged the maintenance of pleasure parks and gardens, because the mild temperatures kept a succession of trees and plants in blossom all year round, brought winter blooms to the rose and almond tree even in northern Italy, and renewed the freshness of the evergreen foliage during the winter rains; hence it rewarded the labor of the cultivator and preserved the beauty of the garden in the cold season. But gardens were the boon of summer. The long, hot, cloudless months made the shelter of vine-grown arbor and cypress avenue a welcome refuge. When the thirsty etesian winds swept down the north or the sirocco from Africa, only the garden kept moist and green and fresh. Eyes tired by a relentless sun and its reflection from the limestone roads rested gratefully upon the dark foliage of laurel or oleander. When the stifling afternoon passed and the people issued from their darkened houses, the garden paths invited to leisurely strolls. In the Paradise legend Adam and Eve heard "The voice of the Lord God walking in the garden in the cool of the day," after the custom of Palestine and other Mediterranean lands.

M. E. Brooke, 1929, p. 43. Our vessel approached St. Thomas before daylight on Thanksgiving Day morning and to those of us who had risen in the small hours and were out on deck at the bow every casual sight and sound had a particularly happy meaning. Each star in the heavens above and every bit of white foam along the ship's sides, where the vessel cut a pathway for itself through the black, placid Caribbean, filled our hearts and souls with the joy of discovery. Every creak of a sailor's rubber boot on deck and each carefully spoken command of the captain on the bridge to his pilot, sufficiently loud to be heard by those on deck near the bow, filled our imagination with thoughts of another captain and another pilot in this same corner of the seas, when grown men were afraid of sea serpents, while sturdy Jack tars were afraid to sail into uncharted seas for fear of falling off the edge of a flat world. The sea had never changed, and ships still mysteriously disappeared and were never heard of again, but now, instead of grown men being afraid of sea serpents, children of four disputed the right to listen through funny little earphones on shipboard through which lounge lizards, hundreds of miles away, transmitted their voices to the uttermost parts of the seas and grown men thought it was good music.

L. H. Halverson, 1930, p. 287. Shut off from the coast by the folded ranges of the Cape, set off on the north from an even less hospitable region by the [mountain] ranges, lies an elevated region of scattered sheep, goat, and ostrich farms, the roughly rectangular Great Karroo. Here great flocks browse upon the scanty herbage, with black herdsmen or woven wire protecting them from marauders.

Preston E. James, 1930, p. 319. Grass! Grass which rises to heights of 12 feet overhead and shuts in the narrow trails; monotonous stretches of grass where nothing can be seen of the country beyond the rank vegetation, with here and there a few low trees all but hidden from view. The prevailing color is brown, a parched dead brown; for this is January, the middle of the dry season, and the vegetable world is at rest. The trees have dropped their leaves. In the scorching heat, tempered by only scattered clouds, everything is withered. Even the rivers have become mere trickles of water. The flood plains stand out as broad sandy flats, glistening white in the sun, marked here and there by stagnant pools covered with green scum, or by reed-filled swamps. Along the river banks are scattered the native villages: small clusters of huts shaped like beehives, and built of grass which blends at this season of the year with the prevailing brown color of the landscape.

Glenn T. Trewartha, 1930, p. 223. Close observation of the hill lands was not possible for want of time. Seen from the plain their forms appear rounded rather than sharp although the slopes are frequently steep. From much of the hill land adjacent to the plain the original tree cover has been removed and these slopes are mantled in coarse wild grasses, practically unused. Farther distant from the basin in the more inaccessible hill country, lumbering is still of consequence. To this the piles of logs at the lower ends of the valleys bear witness.

Adelphia Mitchell, 1931, p. 137. To the stranger who gets his first view of the cove from a little row boat sent out to the steamer, which anchors some distance away from the gently sloping shore, the little box-like homes, the seal skins drying in their frames in the sun, the rude stages and flakes, do not present an inviting picture, and only after one has lived for a time among the people and learned to admire their simplicity, their endurance, their kindliness of heart and courageousness of spirit does he begin to appreciate that the settlement is beautiful in spite of its loneliness and desolation.

Robert H. Forbes, 1933, pp. 246–247. The casual traveler passing by the squalid mud or wattlework villages of these African peoples is likely to underestimate them; but more intimate observation reveals a complex semicivilization. Lacking a script no records are available before Moslem penetration into the interior a few centuries ago. Immense distances and a prevailing condition of intertribal warfare long retarded the penetration of outside influences. Such simple devices as the sail for navigation and the wheel for land transport remained unknown until introduced by Europeans within recent years. Native arts and crafts, therefore, admirably adapted to materials and to natural conditions locally available are certainly in large part indigenous and of ancient origin.

Proficiency in these arts and crafts, furthermore, has prepared the native mind for rapid progress under European tutelage. With instruction excellent blacksmiths, carpenters, cabinetmakers, machinists, masons, and chauffeurs are developed from raw negro material. The native, also, is a good linguist and quick to acquire working knowledge of European languages, business methods, professional occupations, and political organizations—in brief, his brain is not that of an elementary savage but seems in some measure to have been prepared for cultivation and fruitful returns.

R. H. Whitbeck, 1934, pp. 87–88. These are two qualities that a new word ought to have: it should have an agreeable sound, and should define itself. Herein lies the fault of "conurbation." It may stand analysis, but the word does not win my friendship. It sounds too much like the name of an immodest disease. The quality in its favor is cancelled by its disagreeable sound.

E. P. Wheeler, 1935, p. 253. Possibly where time is no object a hand sledge or a toboggan would be even better than a dog team on such journeys in the interior. The problem to be solved in making this choice is whether the comparatively short time that a dog team can travel without a renewal of feed outweighs the slow rate of travel by hand sledge.

Alice Garnett, 1935, p. 601. When on a mountain summit, one seldom tires of watching the movement of sunlight and shadow over the landscape.

Fred B. Kniffen, 1936, p. 179. For Louisiana, house types are an element of culture possessing great diagnostic value in regional differentiation.

John Leighly, 1937, p. 127. A vision of the whole surface of the earth plastered with topographic descriptions—like the luggage of a round-the-world tourist with hotel stickers—is one that must terrify even the most tolerant reader of regional descriptive literature. Among the indefinitely large number of regions that may be subjected to detailed systematic description there must be selection of those that are to be exhibited.

W. Elmer Ekblaw, 1937, p. 215. (''Attributes of Place.'') With discovery and exploration in large part completed, the concern of geography turned to special attributes of place; and out of the mass of knowledge collected and accumulated from world-wide discovery and exploration, a number of new sciences were developed, some of them distinctly related to geography, in that a large part of their fact was more or less resultant upon the study of place, even tho their major content came finally to be quite independent of place relationship.

Ellsworth Huntington, 1938, p. 85. One of the peculiarities of Sweden is the scarcity of villages. As I drove from Stockholm toward the west one summer, three things especially impressed me. The first was the quiet beauty of the agricultural landscape with its green fields, many cattle, and frequent tracts of woodland. The second was the constant succession of lines of boulders and other rough materials marking little moraines where the icesheet temporarily stopped in its last retreat. The third was embodied in the remark of one of my companions—''Do you realize that for the last two hours we have scarcely seen a village?''

Richard Hartshorne, 1939, p. 462. Geography, like history, is so comprehensive in character, that the ideally complete geographer, like the ideally complete historian, would have to know all about every science that has to do with the world, both of nature and of man.

Richard Hartshorne, ibid., p. 640. In order that the vast detail of the knowledge of the world may be simplified, geography seeks to establish generalized pictures of combinations of dissimilar parts of areas that will nevertheless be as nearly correct as the limitations of a generalization permit, and to establish generic concepts of common characteristics of phenomena, or phenomena-complexes that describe with certainty the common characteristics that these features actually possess.

E. H. G. Dolby, 1940, p. 84. Singapore Island, some 24 miles by 14 miles, is shaped like a bat with wings outspread latitudinally.

Felix Howland, 1940, p. 272. Afghanistan has been aptly termed the gateway of India; for from this frontier has come nearly every successful invasion since the first Aryans conquered the country some thirty-five centuries ago. And yet this is not an easy road. The Hindu Kush rise in many places to more than 20,000 feet; and the passes are narrow, rocky defiles little suited to caravans, much less to armies.

Felix Howland, ibid. But all geography is relative. Difficult as the way of the Hindu Kush is, it is by far the easiest means of access to the rich and fertile plains of India. On the east the way is blocked by the soggy, fever-ridden jungles of Burma; to the south is the tempestuous Indian Ocean; on the west the arid wastes of Baluchistan discourage travel; on the north the towering Himalaya form an almost impassable barrier. Only to the northwest are the highways practicable, possessing the five essentials for caravan trade: water, forage, easy gradients, open valleys, and relatively low passes.

Robert S. Platt, 1941, p. 325. Latin America hangs like a pendant—not by land connections with the United States, but by sea connections with countries on both sides of the North Atlantic, in North America and Europe.

Ellsworth Huntington, 1945, p. 174. We have allowed the world to think that there is little or nothing to geography except regional descriptions of the various parts of the earth, a rehearsal of what products result from a given combination of soils, minerals, climate and human activity, and some account of how people live. We talk piously about principles, but those that we mainly stress are extremely elementary and are almost lost under a vast body of facts. I know full well that in saying this I indict myself as well as others, but it is best to face the facts.

Edward A. Ackerman, 1945, p. 138. Every field of professional research and training has to face the problem of what to do with those interested, assiduous members of its fraternity whom I call technicians, for want of a better word. They are the people whose understanding of details much surpasses their abilities in analysis, interpretation, or invention. Almost every reader of professional geographic journals knows what the published results of the technicians amounted to during the past twenty years. A long series of superficial studies describing areas of every size remain as indelible evidence of their efforts. All too frequently "research" merely amounted to one writer's taking in the washing of another. There is exceedingly little of permanent value to be gleaned from many hundreds of pages of these studies. On the contrary, they have a decided negative influence on the reputation of geography, as anyone knows who has discussed the subject with informed non-geographers.

John K. Wright, 1947, p. 7. Unfortunately, this deep-seated distrust of our artistic and poetic impulses too often causes us to repress them and cover them over with incrustations of prosaic matter, and thus to become crusty in our attitude toward anything in the realm of geography that savors of the aesthetic. Like the companions of Ulysses we would row along with ears stopped to the Sirens' song. If a little of its melody penetrate through the stopping, we would try not to let others know. Ulysses himself, however, listened to the Sirens and as a consequence, if one may interpret the matter in a fanciful vein, his whole voyage assumed to him the aura that we sense on reading the Odyssey. Had his companions survived, their accounts of the expedition would have been strictly objective—factual, realistic, but uninspired, and, like some of the geography of today, soon forgotten.

John K. Wright, 1947, p. 10. The realm of geography—geography in the sense of all that has been written and depicted and conceived on the subject—consists of a relatively small core area . . . and a much broader peripheral zone. The core comprises formal studies in geography as such; the periphery includes all of the informal geography contained in non-scientific works—in books of travel, in magazines and newspapers, in many a page of fiction and poetry, and on many a canvas. Although much of this informal geography offers little of value to us some of it shows an insight deep into the heart of the matters with which we are most closely concerned. I venture to think that, of two geographers equally competent in all other respects, the one the better read in the imaginative passages in English literature dealing with the land of Britain could write the better regional geography of that land.

J. Wreford Watson, 1953, p. 313. The discipline had an early start; the ancient world was productive of its cosmogonies and travelogues. Under the Greeks and Egyptians, and particularly under Ptolemy, it even became a systematic, if not scientific, study. The Romans furthered the art of geographical description in various writings about the countries they conquered. But thereafter, geography languished or even lapsed for many centuries.

J. Russell Whitaker, 1954, p. 233. The limited grasp of geographers and the haste to discard recently adopted ideas remind me of an incident told by a missionary who had just returned from Northern Rhodesia. In describing the changes that had resulted there from the introduction of maize, he told of a ripening cornfield that had been ravaged by wild animals. One morning it was found that ears of corn had been plucked and thrown on the ground all along between two of the rows. The next morning two more rows had been treated in the same way. The third night a watcher saw an old baboon come waddling out of one edge of the forest clearing, break off an ear of corn and tuck it under one arm, and then another and put it under the other arm. Then he broke off a third ear and in raising an arm to receive it, dropped the ear being held there and put the more recently plucked one in its place. In this fashion he ambled across the field, breaking off ear after ear, dropping one every time he raised his arm to tuck a fresh one in place. He disappeared into the woods with only two ears of corn.

Carl O. Sauer, 1956, p. 289. The geographer and the geographer-to-be are travellers, vicarious when they must, actual when they may. They are not of the class of tourists who are directed by guide books over the routes of the grand tours to the starred attractions, nor do they lodge at grand hotels. When vacation is found they may pass by the places where they gain the feeling of personal discovery.

Carl O. Sauer, ibid., p. 290. Geography as explanatory description of the earth fixes its attention on a diversity of earth features and compares them as to their distribution. In some manner it is always a reading of the face of the earth. We professionals exist not because men have always needed, gathered, and classified geographic knowledge. The names we apply professionally to the items or forms that we identify and perhaps even to the process we pursue are commonly and properly derived from many vernaculars; we organize them into a vocabulary of wider and clearer intelligibility.

Carl O. Sauer, ibid., p. 298. Why make our regional studies such wooden things which no one may read for the insight and pleasure they give.

Yi-Fu Tuan, 1957, p. 11. Simile and metaphor formed part of the equipment of the early American geologists and geographers in their effort to describe the country vividly prior to analysis. The tradition of vivid description seems to have lost ground in the United States. The modern fashion among professional geographers is to aim at technical descriptions, which, in the extreme case, abstracts the landscape into bare statements and statistics. Of course we need accurate, technical descriptions, but we also need imagery, and words chosen so as to give full appreciation to that object of geographical curiosity, the landscape.

Josip Roglic, 1959, p. 208. Scientific thought and scientific work in small countries, and thus also in Yugoslavia, reflect the achievements abroad and the limitations at home. Interest in geography is very old in Yugoslavia, but geography as a science is relatively young. Local historiography and the teaching of history and even of the Bible supplied information, often erroneous of the lands where the narrated events had taken place. The narrative method is a cumbersome inheritance for geography.

Mary Ursula, 1959, p. 197. From the early shore and river locations, Catholics moved inland with the advent of post roads, public works, canal and railroad building, and the opening up of other economic activities, such as farming, quarrying, lumbering, mining, and manufacturing. Especially notable was the advance of the Irish along the lines of early canals and railroads, for which they became the chief source of labor. Social as well as physical and economic factors were national loyalty, family ties, and love of their religion.

Roland E. Chandon, 1962, p. 72. In classical times, geography was primarily concerned with the description of places. Greek, Roman, and, later, Arab and European geographers observed, measured (in-so-far as they could), and recorded facts about near and distant lands. Notes on the way of life of a community, views on its economic and social conditions, descriptions of the surrounding natural environment—indeed, the entire range of observable phenomena, and a few non-observable items as well—were included in their reports. From this wide-ranging observation and recording of facts, with some attempts at analysis, came the oft-quoted statement that "geography is the mother of all sciences." The close association of geography with astronomy (the latter dealing with spatial relations in the universe, while the former was limited to the earth) was also recognized.

Otis P. Starkey, 1962, p. 261

Dear Kathy,

You ask me, "Can you see much geography from a jet plane?"

You certainly can! But remember that jets usually fly above 30,000 feet, that is nearly 1,000 feet higher than Mount Everest, the world's highest mountain. You must not expect to see everything, for the ground is three to five miles below you. You are too far away to see people, cows, small houses and automobiles, but, if the weather is clear, hills and mountains, rivers and valleys, roads and fields and large buildings stand out sharply.

Yi-Fu Tuan, 1964, p. 439. A problem of geographical description is the difficulty of conveying a visual impression in a sequence of words. In this respect the writer is at a disadvantage when compared with the landscape painter. We can look at a picture as a whole, but we can read only line by line.

Preston E. James, 1967, p. 4. Geographers . . . should not make too much of the supposed distinction between description and explanation.

Edward J. Taaffe, 1974, p. 6. Negative aspects of the integrative view soon began to appear, however. At its worst, it became holistic. Instead of syntheses of interrelated phenomena we had land use inventories or detailed classification systems.

James Parsons, 1977, pp. 14–15. I have found that it pays to keep going back to an area, a people. I have found that significant phenomena or relationships continue to present themselves, things that were at first completely missed or whose significance was not originally apparent. In my experience these linkages are quite accidentally discovered. They are seldom of the sort that can be conjured up in one's study. Repeated visits and growing familiarity with an area continually enlarge one's range of vision and concern. You see, you hear, or read something new, something out of place, something irregular, and you are moved to investigate further. There may be many false leads, many a dead end, but if one's lucky and persists there may be undreamed of rewards awaiting. The experiences, the acquaintances made along the way, can make geography enormously good fun. You learn the ropes, how to get around the libraries and archives, how to use the local newspapers as the rich sources of information they often are, who the people are who know their area, who are at heart the "geographers" with a reservoir of local knowledge and insight above and beyond the norm.

Marvin W. Mikesell, 1978, p. 5. Cultural geographers are attracted to areas and societies that they perceive, rationally or irrationally, to be rich in the traits of culture. It is only in reference to this perception that one can understand why they might find Afghanistan more interesting than Montana or "exotic places" more interesting than "common places."

Marvin W. Mikesell, ibid., p. 8. The setting of the most influential work has been the highlands of New Guinea, where field studies by geographers and anthropologists have provided effective demonstrations of the potential of cultural ecology and hence have set standards that scholars working elsewhere can try to emulate.

Larry Ford and Ernst Griffin, 1979, p. 140. The strongly negative image of the ghetto has been reinforced in the geographical literature dealing with the ghetto.

John Fraser Hart, 1982, p. 26. Regional geography can make an important contribution to a liberal education by helping students to realize that different groups of people in different places really and truly are basically and fundamentally different, and to understand that their behavior is neither more nor less rational than our own, but that it merely is motivated by different values.

John Fraser Hart, ibid., p. 27. Good geographical writing quite definitely is an art, and the highest form of the geographer's art is writing evocative descriptions that facilitate an understanding and an appreciation of places, areas, and regions.

Nicholas Helburn, 1982, p. 447. A couple of years ago I got on the underground at Russell Square in London. It was almost mid-morning, but the sun was just coming up as the train reached Heathrow airport. It was December 21st at 51′N, the beginning of the shortest day of the year. Fifteen hours later, I was whisked off by microbus from Johannesburg to Teyateyanang in Lesotho, 29°S, to experience their shortest night.

I submit that my geography contributed to my understanding and enjoyment not only of the solstice but the vivid contrast between the heart of the empire, the primate city, one day and the periphery, the Basuto enclave within South

Africa, the very next day. Imagine what Suleiman the Magnificent would have given for such a flying carpet.

## REGIONAL UNIQUENESS

Eugene Van Cleef, 1912, p. 401. The City of Duluth holds a rather unique position among cities of the country because of its youth, its rapid growth, and its association with one of the world's most important resources. Its location, topographically, geologically, and climatically, as well as geographically, brings about a combination of conditions which arouses the interest of nearly all its visitors. These conditions are worthy of special consideration because they are so clearly bound up with the geography of not only the locality itself but a vast territory immediately adjacent.

R. H. Whitbeck, 1913, p. 135. Nova Scotians present an example of a vigorous and ambitious people—denied by their geographical limitations adequate employment in business or industrial lines—turning their efforts into intellectual channels and then finding it necessary to seek promotion beyond the bounds of their home province. The chief product of the province has been *men*.

Ellen Churchill Semple, 1913, p. 255. When Japan does anything, the world looks on in an attitude of respectful attention. For Japan has given even the Western world a lesson in efficient methods and effectual results.

R. Malcolm Keir, 1914, p. 769. With Japanese control in Korea, fruit trees might be better protected, but it is not very probable that orchards will soon become a familiar sight, because, in Japan itself, the raising of fruit is a new industry, and the taste and desire for fruit have not yet become general.

C. J. Posey, 1920, pp. 153–154. The great northwest with its invigorating climate and with its resources of fields, forests, and mines furnishes an ample hinterland for the rise of a large urban center whose immediate location is determined by the falls in the great river which taps this hinterland. This site, moreover, is a natural focal point for present day lines of commerce. The rapid and substantial growth of Minneapolis–St. Paul is, therefore, a true geographic expression of the development of the northwest.

Nevin M. Fenneman, 1920, p. 154. What kind of research problems await Geography in the case of a well known state like Missouri?

Albert Perry Brigham, 1924, p. 205. ("Remarks on Geography.") No state more than Virginia, by its plains and mountains, its tidal rivers and bordering seas, exhibits the effects of environment on the course of human events. Near at hand is the ancient mansion of Monticello, which speaks to every true patriot of the man who sent the star of American empire far westward, and whose name enshrines principles that profoundly concern American life for all time to come.

Robert S. Platt, 1928, pp. 81–83. In this article is presented a bit of regional geography dealing with a minute area. The study of Ellison Bay is a primary case, an elementary unit in the science of geography. The area dealt with is not defined arbitrarily, but by the trade bounds of a village community. It is a geographic unit, and regional geography in its lowest terms is the objective of the study. The aim has been to limit the discussion in accordance with the subject discussed. When the investigation began it dealt with an area whose limits were unknown and did not correspond with any measured division; attention was fixed not on certain square miles of land but on the occupancy of land by a certain group of people. In the course of investigation the distribution and range of their activities assumed definite form, a pattern woven to fit the patchwork background of their environment. In the presentation of the study the organized life of this areal unit of human activity provides the theme and limits the discussion.

Wallace W. Atwood, 1928, p. 269. There is ample evidence of suspicion, due to ignorance, directing the thoughts of American people when they consider the actions of peoples in other lands.

L. Dudley Stamp, 1930, p. 86. In many respects Burma is an anomaly.

Harold S. Kemp, 1930, p. 224. It is the purpose of this paper, not to prove Brittany a surviving island of the picturesque, but to demonstrate (what must be true in the case of any such survival in the midst of a modern world) that its picturesqueness is the product—the reflection—of such prosaic elements as climate, rock structure, location. In other words, it is a matter of geography; in this case, social geography.

Harold S. Kemp, ibid., pp. 228–229. In the rear of the church . . . stands another Breton expression of environment—the . . . bone-house. In a region of so little soil, the burying of the dead long ago threatened to become a grief-ending worry and expense to the living; hence, after a few years of mourning, the bones were dug up from their snug but scanty resting place and dumped, with more or less ceremony, into the bone-house, which, occupying a modicum of land, and made of age-resisting granite, solves the problem very well indeed.

Harold S. Kemp, ibid., p. 229. Brittany's remoteness lies chiefly in the fact, common to peninsulas, that the province leads nowhere.

Oliver Richetson, Jr., and A. V. Kidder, 1930, p. 191. At 1:42 P.M. the main escarpment, running northwest and southeast, which had been noted the day before, was clearly visible on the horizon far to the west.

Ralph H. Brown, 1930, p. 271. The city of Boulder, as a quality of its site astride . . . Boulder Creek at its canyon-like exit from the adjacent foothills of the Front Range of the Rocky Mountains, presents within its immediate compass natural conditions of remarkable diversity. In this respect the city is unique among the several urban centers of the piedmont of northern Colorado.

K. J. G. Sundaram, 1931, p. 49. There are about fifteen times as many villages as there are automobiles in India.

Robert S. Platt, 1931, p. 52. ("Urban Field Study.") Marquette has a population of 15,000 and property assessed at $11,000,000. Yearly it receives 4,000,000 tons of commodities and sends out a similar outflow. This concentration of dense population, valuable buildings and restless movement, in its external relations and its internal structure offers a subject for geographic interpretation.

Billie L. Dickson, 1931, p. 160. You may copy this paragraph in your geography note books. In the present site of Spokane the early settlers recognized a natural railroad center. The falls of the Spokane river furnish power to manufacture the wheat, lumber, and other raw materials from the surrounding country. The attractive scenery, healthful climate and many different ways of earning a living, have all helped to make Spokane a growing and prosperous city.

Daniel Bergsmark, 1931, p. 111 Located immediately east of Cincinnati, Clermont County has a favorable situation for economic growth. It extends 33 miles from north to south, 16 miles from west to east, and embraces 465 square miles. Bounded on the south by the Ohio River, the county contributes its mite [sic] to the stream of traffic on this major inland waterway. With a frost-free season of 173 to 200 days and mean annual precipitation of 40 inches, it lies within the Corn and Winter Wheat Region of North America, where corn and winter wheat are the dominant grain crops in the rotation. Its location in the Till Plain along the southern margin of the old eroded glacial drift is reflected in a paucity of moraines and lakes within its borders. But this plain has been dissected in some places, and the larger streams have cut valleys from 200 to 400 feet deep, leaving a land surface composed of level and rolling interstream areas, valley slopes, and valley bottoms, with their different environmental conditions for agricultural production.

Robert S. Platt, 1931, p. 215. ("Pirovano.") This study has to do with a tract of land in central Argentina—not a community, not a distinct district, not even a well-defined territory, but an undifferentiated fragment of a large region. The choice of the area of Pirovano for study does not represent a preference for such a fragment instead of a complete unit of human occupancy, but does represent exigencies of field work in a large uninterrupted region of great complexity.

R. H. Whitbeck, 1932, p. 27. Like most regions, Jamaica has both its geographical assets and its liabilities.

Frank E. Williams, 1932, p. 85. The problem of crossing the Andes has always been a serious one.

Richard Hartshorne, 1932, p. 431. The cultural landscape of the St. Paul–Minneapolis urban district presents a picture of almost unique character.

Stephen Sargent Visher, 1932, p. 288. Nearly every mile of the world's more than 222,000 miles of international boundary was located where it is as a result of earnest efforts on the part of leaders who had certain objectives in mind.

Mark Jefferson, 1933, p. 90. ("Great Cities.") The American public takes lively interest in the size of the city of New York. Is it larger than London? An odd sort of patriotic bias tinges the question. It is difficult to estimate the height of the atmosphere because it has no top—the air peters out. In the same way a great city thins out and comes to no definite limits.

W. M. Davis, 1934, p. 11. ("Long Beach.") Many persons moved away from San Francisco shortly after the earthquake and conflagration of 1906, but in spite of their withdrawal the city is now larger than ever before.

Joseph Earle Spencer, 1935, p. 353. [In China] salt and iron became government monopolies, more or less, several hundred years before the opening of the Christian era, and salt has ever since provided the central authority, the province, the military feudal lord, the outlaw holder of a transportation route, the local political unit, and, of late years, the national government with a steady source of revenue.

Robert Burnett Hall, 1935, p. 127. One reason for the lack of progress is that the region has too often been regarded as simply a convenient device by which to limit and characterize the worker's study and has not consciously been integrated into the world order of things. Another reason is the apparently irresistible temptation to stress the uniqueness of the particular region and the forms and relationships within it at the expense of general and character description.

Robert S. Platt, 1935, p. 153. Last year . . . I was called the most near-sighted of the microbes examining the earth's surface in microscopic geography.

Robert S. Platt, 1936, p. 154. Curacao is known as an entrepôt and as a liqueur.

Lawrence Martin, 1938, p. 303. It seems incredible, but the date and the occasion of the naming of Mount Washington, New Hampshire, have not been established.

Ben F. Lemert and Rose V. Lemert, 1938, p. 188. Popocatepetl is one of the most famous mountains of North America. Who does not remember trying to spell its name during school days? The Mexicans who live on it do not even try. They just call it Popo.

Edward A. Ackerman, 1938, p. 253. Albania has been described as the Switzerland of the Balkans. The metaphor is not as absurd as it at first seems, for Albania, like Switzerland, is a small mountainous country whose rugged relief and focal position have always preserved for it a measure of independence.

Mark Jefferson, 1939, p. 226. Metropolitan London is not merely the largest city in the world; it is more than seven times as large as Britain's second city,

Liverpool, and thereby stands out alone in a different order of magnitude and significance from those of all other cities in the country. It is the primate city of the United Kingdom.

Robert S. Platt, 1939, p. 105. Among the several types of recent geographic work the following study may be put categorically in its place. It is regional geography and not systematic, since it deals with the complex of phenomena in an area. It is pure geography and not applied, since it is directed toward understanding phenomena as they are without an ulterior motive of reform. It is academic but not pedagogical.

It is human geography and it is anthropocentric, but it is not anthropogeography as narrowly defined and not human ecology. The viewpoint is not that of environmental determinism or controls or influences or responses or even adjustments or relationships.

It is not Geopolitik, though it has a touch of political geography, as well as economic and ethnographic. It is not culture-form geography, though it takes account of the topography of art. It is not genetic morphology. But it is concerned with landscape, with areal association of observable phenomena, with pattern of terrane occupance. Regional units both of static areal homogeneity and of dynamic organization are taken into consideration.

The study is chorologic and not cosmologic. Within chorology it is microchoric rather than macrochoric. It belongs essentially in the field of microgeography. The study deals primarily not with all of British Guiana but with two plantations on the coast, occupying less than 1/10 of 1 percent of the area of the colony.

Robert S. Platt, ibid., p. 125. Microgeography is a logical development in modern method, begun before the World War and still evolving—not a fad of the 1920s now abandoned. Its leaders have been familiar with post- and pre-war geography, geological and botanical field techniques and military mapping, and have been conscious of defects as well as contributions of both environmentalism and ecology. The movement is easily understood as a rational and timely drive against the limitations of armchair compilation from promiscuous data, or subjective impressions from casual travel, and of environmental theory not founded on observed data.

The occasion has called for objective data, congenitally geographical, specifically located in areal association. Field work is a prerequisite, and in the field is an old obstruction: the geographer's dilemma in trying to comprehend large regions while seeing at once only a small area. In the field all geographers are "microscopic."

Richard Hartshorne, 1939, p. 438. Geography has at least one individual, unitary, concrete object of study, namely the whole world.

Richard Hartshorne, ibid., pp. 540–541. Finally, there is one factor of great importance in regional geography which we cannot classify, namely, the factor of relative location of one place with reference to another. An essential feature of any area is its location with reference to the other areas of the world, both near and remote, both land and sea. Clearly this cannot be classified in a finite number of systems. In terms of its relative location, or *locus*, each area is unique; the facts cannot even be adequately expressed in words but can only be shown on a map—or rather, on a globe. Once this factor is introduced, therefore, we must shift from systems that classify areas to a system that recognizes specific areas as they are actually located.

Richard Hartshorne, ibid., p. 625. Regional geography, we conclude, is literally what its title expresses: the description of the earth by portions of its surface. Like history, in the more common sense of periodic history, it is essentially a descriptive science concerned with the description and interpretation of unique cases, from which no scientific laws can be evolved.

Richard Hartshorne, ibid., p. 644. The findings of regional geography, though they include interpretations of details, are in large part descriptive. The discovery, analysis and synthesis of the unique is not to be dismissed as "mere description"; on the contrary, it represents an essential function of science, and the only function that it can perform in studying the unique. To know and understand fully the character of the unique is to know it completely; no universals need be evolved, other than the general law of geography that all its areas are unique.

J. Granville Jensen, 1945, p. 162. One way to make geography real is to turn our attention for at least part of a semester to the home environment.

Robert S. Platt, 1946, p. 1. I must admit explicitly a fact which has always been implicit in microgeography: that the little spot of which I speak is included only because of its larger significance in relation to the world and as it contains the world—not merely as a nerve cell in world order, like "the flower in the crannied wall" which holds the secret of life and is affiliated with the universe.

Robert S. Platt, ibid., p. 2. Problems of our time involve the interlocking of human life over the whole earth, and simultaneously the localization and even to some degree the isolation of life in every individual place on the earth. In one square mile there is evidence of these antithetical aspects of life—world unity and disunity, coherence and incoherence, attachment and detachment, conjunction and separation, interdependence and independence—opposite but coincidental aspects of life to be met and dealt with willy nilly by everyone who lives on the earth, and subject to understanding by geographers if understandable by anyone at all.

Preston E. James, 1948, p. 271. At the beginning of any serious research study it is essential that the objectives be clearly and unambiguously stated.

Preston E. James, ibid., p. 273. Geographers are always dealing with categories of things. No two points on the face of the earth are alike; yet differences, similarities, and relationships are identified and analyzed in terms of categories in which similar things, but not identical things, are grouped together. Since the selection of categories can determine the kinds of relationships identified, it is highly important that they be selected in such a way that they can be evaluated by one's professional colleagues. This requires a statement of the objectives, for only in terms of objectives can categories be evaluated.

Raymond E. Crist, 1952, p. 338. Some of the best-known and tastiest jellies and pastes (*pastas*) of tropical America are made from the guava, or *guayaba*, the yellow, crabapple-like fruit of the *guayabo* tree (*Psidium guajava L.*), often called the *guayabo casero*. The dessert *de riguer* in many homes as well as in the hotels is *pasta de guayaba* with white cheese. In the tiny hotel in the little inland town of Jequié, Brazil, this dessert was known as "the three-hundred-and-sixty-five" because it was the one and only dessert served every day of the year at all meals except breakfast.

Fred K. Schaefer, 1953, p. 227. Geographers writing on the scope and nature of geography often begin quite apologetically as if they had to justify its very existence. And strangely, or perhaps, psychologically speaking, not so strangely, they go on claiming too much. In such writings geography, together with history, emerges as *the* "integrating science," completely different from other disciplines, whose unique importance finds its expression in the special methods which it must use to reach its profound results.

Fred K. Schaefer, ibid., p. 231. We have seen that there is a whole group of ideas which are variations of a common theme: geography is quite different from all the other sciences, methodologically unique, as it were. Influential and persistent as this position is in its several variations, it deserves a name of its own. I shall call it *exceptionalism*.

Fred K. Schaefer, ibid., p. 238. The main difficulty of the uniqueness argument is that . . . it proves too much. Are there really two stones completely alike in all minute details of shape, color, and chemical composition? Yet, Galileo's law of falling bodies holds equally for both. Similarly, limited as our present psychological knowledge is, it seems safe to say that no two people would register identical scores on all tests as yet devised. Does it follow that our psychologists have so far discovered not a single law? What it all comes down to is a matter of degree. In the physical sciences we have succeeded in discovering a set of variables such that if two objects or situations, no matter how different they are in other respects, agree in these variables or indices, then their future with respect to these indices will be the same and predictable. To what extent and how soon any other field will reach a state as satisfactory as this is a matter of fact, to be decided by trial and error, not to be prejudged by pseudomethodological argument.

Fred K. Schaefer, ibid., p. 240. Geography according to Hartshorne is essentially idiographic.

J. Russell Whitaker, 1954, p. 233. If one has studied and digested the thinking of geographers for a half century, one surely has sufficient wisdom to see that both "distributions" and "relationships" matter; that changing conditions of the physical earth do "influence" man; that man does make "adjustments" to earth conditions; that the "content of area" does present a challenge to geographers; that, however much geographers are interested in the whole earth, they find it convenient to divide it into parts; that, however carefully they center their efforts on a subdivision of the earth, they fall short of their goal if they ignore its spatial relations with the rest of the earth; that geographers should have room in their thinking for both the uniqueness of every place and the oneness of the world. And so I might go on to identify values in many other ideas about which geographers have debated during the last fifty years.

Richard Hartshorne, 1955, p. 206. The primary purpose of this paper is to provide correction to false representations and accusations in an article published in the *Annals* of 1953. The false statements are examined and corrected in the main body of the paper. Those who did not read the original article or recognized its unreliability may well save time by ignoring both this re-examination and the original article.

Benjamin E. Thomas, 1956, p. 434. Few towns of equal size have captured the imagination and endure in geographical folklore as Timbuktu has. The name has lasted for centuries as a symbol of remoteness.

G. H. T. Kimble, 1962, p. 1. Many people still think of the Geographer (if they think of him at all) as a dealer in terrestrial bric-a-brac, whose sole social function is to provide other people with the answers to quiz questions.

Willis B. Merriam, 1962, p. 68. Almost everyone appreciates steak smothered in mushrooms, cream of mushroom soup, or the several mushroom sauces that dress a pizza or a spaghetti dish. Few have any notion whatever about how mushrooms are grown, where they are raised, or the technical intensity of their cultivation.

The area within a radius of 25 or 30 miles of Kennett Square, in Chester County, Pennsylvania, is the nation's and probably the world's leading center of commercial mushroom production.

Fred Lukerman, 1962, p. 348. If geography is the study of places, it follows from logical considerations as well as experience that the primary geographic concern is not with the distribution of things (natural phenomena, man, his works) on the earth's surface, but with localities, parts of the earth's surface. A place is not a thing but a fact—a matter of human consciousness, a consciousness about things, a matter of experience.

Edward A. Ackerman, 1963, p. 438. Selection of a research problem at random . . . risks triviality, even though it may be entirely "geographic" in conception.

Peter W. Lewis, 1965, p. 27. Logically, everything is unique. But similarities can be found among unique things and general statements can be made from these similarities. These statements are then expressed as formal laws.

William Bunge, 1966, p. 375. Locations are not unique. They are all located relative to each other. Locations do vary in some degree by all the members of the universe. All the locations on the earth's surface do possess in some degree of similarity, for example, the location 10 Downing Street. The locations at 10 Downing Street fully possess 10 Downing Streetism. The locations in the immediate vicinity possess 10 Downing Streetism to a high degree. The locations in Madison, Wisconsin, possess it to a very low degree but in still higher degree than do locations in Iowa City; that is, some locations are much more similar than others.

William Bunge, ibid. It is true no location is exactly like any other, but then no two anythings in the real world are exactly alike. To admit that no two objects are identical, that is, have everything in common, in no way contradicts the statement that these objects might have much in common. Not to be identical does not imply being uniquely different. Snowflakes are not identical but they share many features. The same is true of fingerprints, individual human personalities, and locations. All are classifiable. Why are geographers so stuck over this simple logical point?

Kevin C. Kearns, 1968, p. 167. The pleasure a visitor derives from city parks is not a fleeting one. No matter how interested he has been in other sights, what lingers longest in his memory is the recollection of having passed some pleasant hours in a region of rest that was far enough removed from the turmoil of life to make him forget his worries.

Martinez Estrada Ezequiel, 1971, p. 143. The men who inhabit the southern extremity of the globe live in Vshnaia. It is a penitentary, the heavens are its closest neighbor.

James Parsons, 1977, p. 15. For me, at least, it all comes back to place, to area as the integrating concept, to the supremacy of observed geographic data over any pyramid of deductions or formal theories, however powerful the apparatus brought to bear. Geography by its nature offers the promise of carrying us beyond generalizations to more exact knowledge of the interaction of man and environment. Diversity, we are learning, seems to be a major stabilizing force in the natural order, and probably in the social order, too. It is something to analyze and understand, not something to be gotten rid of. Interest in the unique, moreover, may be the single most powerful recruiting agent we have for bringing students into geography. We must be aware of those who would brainwash us into thinking it is something shameful or secondary, that only generalizations or theory are wholly respectable.

John Fraser Hart, 1982, p. 14. I think geographers can spare themselves considerable unhappiness and rejection if they will refrain from addressing scholars who are not interested in what they have to say and who are not competent to evaluate it.

Yi-Fu Tuan, 1984, p. 245. Last year continuity and discontinuity were much on my mind. As has been a habit of mine, I transposed the personal problem with its inevitable ambiguities onto the bright-lit stage of an impersonal question. Why make another move? Why cut the roots that one has established, perhaps for the first time in one's life? One answer is that I have always felt that being in place—being rooted—is an illusion. I have always been haunted by the idea of departure both of the glad, self-initiated and of the wrenching, unwilling kind. We are all more or less aware of a final departure that awaits us, but we are less aware of it when we are in the presence of friends and good books and are comforted by their projection of continuity and of reverberating meaning.

## NATURAL REGIONS AND REGIONALIZATION

Wolfgang L. G. Joerg, 1914, p. 55. With the recognition of regional geography as the ultimate goal and highest expression of geographical research, which has come with the modern development of our science, there has been a marked increase in the attention devoted to this branch of the subject and to its methods. One of the most important questions of method is that which refers to the unit of investigation. Economy of presentation and sound geographical reasoning both demand that such units be as homogeneous as possible. Inasmuch as this requirement is but rarely fulfilled by artificial units, such as political divisions, and as geographical phenomena appertain to the physical world, modern workers have urged and practiced the use of natural regions as the fundamental unit of geographical investigation.

Wolfgang L. G. Joerg, ibid., pp. 56–57. A natural region may be defined as any portion of the earth's surface whose physical conditions are homogeneous. This exclusion of the human element does not imply any disparagement of anthropogeography. On the contrary, man's activities will be interpreted all the better the more clearly his environment is recognized. In fact, it may be said that the recognition of the desirability of subdividing the earth's surface into natural regions is a direct consequence of the development of regional anthropogeography.

Charles R. Dryer, 1915, p. 122. The final cause of a natural region, in the Aristotelian sense of an end to be realized, is economic. If the industries and occupations by which men get or can get a living in a given natural region are not distinctive and different from those of the surrounding territory, that region, as delimited, lacks essential unity and utility. Here the geographer may be sure of his ground. By confronting every natural complex with an economic complex it will be compelled to reveal its essential geographic significance. In *economic influence* may be found a standard for estimating the relative value of the different factors which make up a composite natural region.

Mark Jefferson, 1915, p. 30. It is time to give a little more attention to the *composite* cities. All studies of cities and city-growth involve some consideration of the meaning given to the word "city" in each case.

Helen Gross Thomas, 1920, p. 253. Regional geography is based upon the divisions of the continents into natural regions. A natural region is an area within which the geographic conditions (primarily topography and climate) are broadly similar, giving rise, perforce, to certain occupations and habits of life among people.

Charles Redway Dryer, 1920, p. 11. Regional geography is now being placed on a scientific basis by the adoption of natural regions as units.

R. H. Whitbeck, 1923, p. 86. For some years past certain geographers have from time to time advocated the division of the continents into what they have called natural regions, which regions should become the units of study in geography. This idea of natural regions has been more in evidence in Europe than in America; both French and English geographers have devoted considerable time to efforts in this direction. A large number of maps have been constructed in which the world or parts of it have been divided into so-called natural regions.

Sten De Greer, 1923, p. 497. Most geographers now agree that cities should be studied systematically as important, complicated, and interesting geographical regions albeit small in areal extent. Such treatment is especially called for in the case of the large cities, where the urban regional type is most distinctive and where the most complete subdivision of function is exhibited, that is where subregions can be identified.

Preston E. James and R. Burnett Hall, 1924, p. 288. Environment and man are the two major elements in a geographic study. Geography interprets the manifold relationships which exist between man and the world in which he lives. As the number of such interpretations increases, certain sections of the earth appear in which the environmental factors are somewhat similar, and as a result, in which man is engaged in somewhat similar activities. Probably no two localities have environments which are exactly alike, but sufficient resemblance exists between neighboring localities so that they can be grouped together into what is called a region. The definition of such a region might be: A considerable part of the earth's surface thruout which there are similar dominant conditions of physical environment, and consequently similar dominant human activities. The physical environment and the human activities are the two major elements in this concept.

D. S. Whittlesey, 1925, p. 49. A geographic region expresses unity in relationship between life conditions and conditions of the natural environment. It is allied to natural regions, and to regions of human activities, but it is not identical with either. A major geographic region is the largest area in which geographic unity appears.

Melvin R. Gilmore, 1927, p. 161. The Missouri River has been a powerful factor in shaping the activities and destinies of the Indian natives which have occupied its basin through many centuries past. Its course has been the great highway between north and south, and its tributaries have been the lines of travel between east and west. The course of this river has directed and facilitated the streams of migration of human population, the exchange of useful material commodities, and the flow and interchange of intellectual goods in human culture.

Millicent Todd Bingham, 1928, p. 24. The city is not only a legitimate subject for geographical analysis, it is the epitome of regional geography, its elements being physical factors such as topography, soil, climates, waterways, and human factors such as industrial possibilities and routes of communication.

Mark Jefferson, 1928, p. 220. Mediaeval towns built walls to keep their enemies out and for the most part failed. We have not learned to build the best possible road out to our city limits and put up the sign there "Welcome." We cannot keep people out, it appears, and it is very much more to our advantage to have them come in.

A. E. Parkins, 1931, p. 1. A . . . postulate that we accept is that there are types of human behavior, or adjustments, and that the physical environmental complexes of the earth may be grouped into regions, thus giving us regional behaviorism, or regional geography.

Derwent Whittlesey, 1932, p. 84. The classic example of the ideal mountain boundary is the Pyrenees.

Richard Hartshorne, 1933, p. 196. The pursuit of such vague concepts as "natural boundaries," a term seldom defined and usually meaning something different to each writer . . . , should be banned from scientific literature.

V. C. Finch, 1934, pp. 114–115. A geographical region, or even an arbitrarily-chosen portion of the earth's surface, may be thought of as having some of the qualities of a human being. It is a thing both physical and cultural, not wholly physical, not wholly cultural, but with physical and cultural elements so interwoven as to give individualism to the organism. Like human beings, geographical regions do not always exhibit a nice balance in the development of both sides of their natures. Just as the savage man is dominantly a physical creature, his cultural nature yet to be developed, so some regions of sparse or recent human occupance remain dominantly natural. At the opposite human extreme is the true esthete in whom the instincts of savage man are almost completely subordinated to a set of highly cultivated values, aspirations and inhibitions. Some regions are like him, particularly some urban regions. In them natural earth has been all but obliterated, hills cut down, valleys filled, drainage reversed and soils remade or buried under vast expanses of construction and pavements. Even the rigors of primeval climate are to a great degree offset, if not overcome, and the populace lives and works with but little thought of the natural forces to which in other regions, they would need to shape the manner of their existence. Between the two extremes in human culture are the great majority of mankind.

Wellington D. Jones, 1934, p. 106. The determination of regional, sub-regional, district, and locality boundaries is a perplexing problem because in so many cases one "essential homogeneity" gradually gives way to another with no sharp line separating the two. This problem is solved by some geographers . . . by refusing to draw boundaries.

A. E. Parkins, 1934, p. 223. There is quite general agreement that scientific technique demands that the earth may be considered as a mosaic of areas and regions, scores and hundreds in number; but hundreds of areas and regions may be grouped into a smaller number of units on the basis of one, two, or several common characteristics. That is the concept of regional geography accepted by all or nearly all geographers in America. Some geographers would stress the human activities in the region; others the cultural region which is the result of human activities. Some would emphasize man's part in the geographic region. Others would lay stress on the natural features as being as important as human items.

Robert Burnett Hall, 1935, p. 122. The major contribution of geography to the general field of science is the recognition, first of the ever-varying aspect of the land, and secondly, that, in spite of this variation, the land tends to be divided into areas of more or less similarity. Such areas we call regions.

G. T. Renner, 1935, p. 137. It has been rather generally accepted among American geographers that a region is an area which is homogeneous enough in its physical character to possess either actual or potential unity in its cultural aspects.

K. C. McMurry, 1936, p. 94. The various statements of the definition of geography which are in wide use today all agree on two points: (1) the necessity of focusing attention on the region or area as a unit of investigation, and (2) in the inclusion of a very wide group of phenomena—physical, economic and social—in so far as they are associated within the areal complex.

G. Donald Hudson, 1936, p. 110. The unit area method of land classification as evolved contributes both to the field of geography and to the field of planning. In geography it affords another tool of investigation—one that partially fills, at least, the gap between methods of detailed field analysis and methods of reconnaissance. In planning, its use yields results that are vital to the development of an effective land plan.

Derwent Whittlesey, 1936, p. 199. Of all the modes of using land, raising crops and livestock covers the most space and is the most easily observed in the field. It therefore lends itself better than others to geographic classification—that is to say areal analysis and synthesis.

Richard Hartshorne, 1937, p. 161. Countless articles have been written to enlist our sympathies in Germany's claim to "more equitable boundaries." It is not the geographer's place to judge of rights and justice, but he is interested in analyzing the area facts and relationships involved in the problems of the Corridor. This indeed is an essential, tho incomplete, basis on which the individual may form his own judgement [sic].

W. Elmer Ekblaw, 1937, p. 219. ("Attributes of Peace.") Regional geography, in America at present the most popular approach or method in the teaching of geography, possesses distinct advantages for the use of place discipline. The region is merely a collection of places grouped upon the bases of some convenient, distinctive, or peculiarly useful attribute, or composite of attributes, which the places possess in common. The regional map constitutes a classification of attributes that can be graphically portrayed. The regional approach holds facts and principles strictly to geographic discipline; but it possesses the inherent danger that in the minds of the students, and perhaps of the teacher as well, the region assumes a distinct natural entity, instead of remaining merely a classification represented graphically upon a map.

Stephen S. Visher, 1938, p. 301. ("International Boundaries.") Omitting small islands and minor irregularities of the coastal and river boundaries, the international boundaries of the world today have a total length of about 222,200 miles of which nearly a half is coastal, about a fifth is mountain, another fifth is artificial, and about one eighth follows rivers.

V. C. Finch, 1939, p. 10. No geographer is unaware that chorology, in the sense of a complete, wholly consistent, and strictly rational regionalism is difficult if not impossible.

V. C. Finch, ibid., p. 11. Can it be shown whether regions, as the chorologist treats them, have any reality? Can geographers who continue the attempt to delimit, describe and interpret earth features by regions have any basis for scholarly satisfaction outside the artistic method which the skeptics admit to respectability; outside the treatment of process and change which some would make the organizing principle of regions? Can any clear answer be given to the question, recently asked in a meeting of geographers, whether the regionalist, even in the classroom, is scientist or mountebank?

Edward Ullman, 1939, p. 291. For more than 250 years the eastern Rhode Island–Massachusetts boundary was a subject of dispute. At times both states angrily threatened to call out the militia, and large sums of money had been expended in litigation before final settlement was made in 1899.

Richard Hartshorne, 1939, p. 537. Although the earth surface—the world—in terms of its geographic features, is not divided into distinct areal parts, the fundamental function of geography—the understanding of the differences between different areas—requires the geographer to divide the world arbitrarily into area parts. A general principle of any science requires that its material is not to be left in patches, but is to be organized into a logical system of knowledge.

Richard Hartshorne, ibid., p. 641. The problem of dividing the world, or any part of it, into subdivisions in which to focus the study of areas, is the most difficult problem of organization in regional geography.

Derwent Whittlesey, 1943, p. 1. The geographer views the earth as a mosaic of regions. These may be large or small, but every region is marked by a twofold unity. Its physical nature is broadly uniform and the mode of life practiced by its inhabitants is coherent thruout the area.

Stephen B. Jones, 1943, p. 99. Cases of international discord of serious nature have been caused by slight and unintentional ambiguities in the description of boundaries in formal documents. These flaws may be due to unfamiliarity with the peculiarities of the geographical features, human or natural, along which the boundary extends, or to lack of knowledge of the pitfalls in boundary description. The wisest settlement of a territorial dispute may lead to friction if the description of the boundary in the treaty or award does not correspond to geographical realities. Words that seem simple and straightforward may prove stumbling-blocks when surveyors endeavor to demarcate the line upon the ground. A border officer in future years may wrestle with some problem that began, almost literally, in a slip of the pen.

Robert S. Platt, 1943, p. 230. There are regional differences and separations on the earth but no totally distinct regions with absolute boundaries. Regions are relative generalizations, defined according to selected criteria, constructed subjectively or organized temporally for special purposes.

S. S. Visher, 1948, p. 282. Despite Indiana's small area (about 36,000 sq. miles) and slight range in elevation (313 to 1285 feet, with about four-fifths within 650 to 1,000 feet), there is sufficient areal contrast so that a number of regions are generally recognized. Detailed studies divide the state into four to twelve regions and there are logical bases for additional sub-divisions.

Werner J. Cahnman, 1948, pp. 234–235. In the language of geography, area studies are studies of "specific regions" rather than of "generic regions." Studies of generic regions attempt to classify areas "regardless of location and association in space" while studies of specific regions aim at an "areal division of the world into major parts, these each subdivided into subdivisions which are contiguous and together form an associated whole." Generic regions, such as areas of high or low precipitation, high or low population density, and the like, are characterized by one selected factor and studied under the aspect of generalization and law while specific regions, such as communities, nations, and culture areas, are characterized by a multiplicity of interrelated factors and studied under the aspect of wholeness and uniqueness. The study of generic regions reveals the similarity of widely separated areas; the study of specific regions, however, shows that this similarity is fundamentally incomplete, since the location and association of an area is one of its important inherent characteristics which may modify other inherent characteristics to a considerable degree. Specific regions, under circumstances which will be discussed presently, may be comprehended as "social fields." In this case, they transcend the boundaries which are drawn by geography and partake of the character of all the social sciences.

Lyle E. Gibson, 1948, p. 244 and p. 270. The line along which . . . economies meet is defined as a regional margin, the place (zone) of contact between two types of economic occupance. . . . The margin in this study is a meeting place of the cheese and meat animal economies, and many individual farmers produce both milk for cheese and meat animals. The majority of farms, however, emphasize one type of production or the other. In other words, it is quite probable that the inter-regional spaces delimited as general and mixed farming in previous studies may be reduced considerably by more detailed study.

G. Donald Hudson, 1951, p. 110. The region as a concept is vital to geography. It is a simple concept. We should keep it so. When applied in geography, the region as a concept becomes essentially a device—a contrivance—a scheme. So vital is it that it attracts our attention. As a result we become, from time to time, concerned with the details of its mechanisms. So concerned, we are in danger of creating of it a monstrous thing, too complex and formidable to be used for the purpose it was designed to serve. Thus we lose its utility.

G. Donald Hudson, ibid., pp. 110–111. The region in geography has but one purpose. Geography deals with phenomena that have the quality of distribution. Findings in geography with respect to these phenomena must be presented so that their true distributional quality is revealed. Understandings derived from these findings must be similarly demonstrated. In its application, the region as a concept becomes the device by which these requirements are met. In this sense, the region as a concept is a guiding principle in geographic research, but in its ultimate application it is not a tool of research. It becomes a device of presentation and demonstration.

Preston E. James, 1952, p. 199. The region is a geographic generalization.

Edward L. Ullman, 1953, p. 58. The geographer should be able to contribute basically to the regional delimitation problems of the area under study, whether for the whole area in relation to others, or for parts of the area. This is merely a means to the ends of area study, but one on which much heat and little light is often generated.

D. Whittlesey, 1954, p. 30. . . . the committee came to see the region as a device for selecting and studying areal groupings of the complex phenomena found on the earth. Any segment or portion of the earth surface is a region if it is homogeneous in terms of such an areal grouping. Its homogeneity is determined by criteria formulated for the purpose of sorting from the whole range of earth phenomena the items required to express or illuminate a particular grouping, areally cohesive. So defined a region is not an object, either self-determined or nature-given. It is an intellectual concept, an entity for the purpose of thought, created by the selection of certain features that are relevant to an areal interest or problem and by the disregard of all features that are considered to be irrelevant.

Carl O. Sauer, 1956, p. 297. I know of no general or inclusive descriptive system for regional study that has the promise of a real taxonomy.

Carl O. Sauer, ibid., p. 298. The "unit area" scheme of mapping may be a useful cataloguing device like the decimal systems of librarians, though I doubt it, but as a means of research I should place it below almost any other expenditure of energy.

Rollin S. Atwood, 1957, p. 151. The region was not chosen on the basis of physiography, climatology, soils, or vegetation. It was chosen as an area where a serious problem facing the economic stability of Chile could be effectively attacked with the facilities available.

Leonard Zobler, 1958, p. 140. Regions are areal systems based on levels of similarities and differences in spatially distributed traits. Regional boundary lines are conceptual areal frames fitted to the earth's surface within which landscape features are sorted. It is the nonhomogeneous distribution of these features which identifies the peculiar properties of respective regions. Each spatial unit is internally homogeneous but differs from other members of the system.

Richard Hartshorne, 1958, p. 268. ("What Do We Mean.") Empirical examination of our literature shows that geographers use the same word "region" to express any one of three different concepts: (1) A "region," in the general sense, is simply a particular piece of area which the writer regards as being in some way distinctive from other areas. (2) A homogeneous, or uniform, or formal region is an area within which the variations of one or more selected features each fall within a certain narrow range. (3) A region of coherent organization, or a nodal or functional region, is an area in which one or more selected phenomena of movement connect the diverse localities within it into a functionally organized unit.

David Grigg, 1965, p. 465. It has often been noticed that whereas many regional systems have been devised by geographers there have been relatively few attempts to suggest any principles of regional divisions.

Richard Hartshorne, 1965, p. 619. Theoretical discussions of the nature of regions commonly omit a major category of real units, namely, sections of area instituted by human agency as distinctive and discrete units limiting the areal extent of operation of certain functions.

Robert B. McNee, 1966, p. 64. At least three values traditionally held by geographers include (1) a traditional preference for the holistic or comprehensive approach, (2) a traditional emphasis on direct observation, and (3) an appreciation of the aesthetic in maps. Though found at every scale, holism is most marked in their traditional yearning for a comprehension of the whole planet as a single system or series of sub-systems.

James R. McDonald, 1966, p. 528. For geographers seek regions as alpinists seek unconquered peaks: "because they are there," and the fact that their existence is in the correlative nature and ecologic sensitivity of the geographer, rather than in a visible object of granite and ice, detracts not at all from the analogy.

R. J. Johnston, 1968, p. 575. Problems of classification are particularly severe in geography because of the discipline's interests in spatial patterns.

William W. Bunge, 1973, p. 275. ("Geography of Human Survival.") The biological crisis of existence in which the species finds itself is clearly demonstrated in geography. The spread of mankind and machinekind explains the spread of speciecide. Geobiological revolutions are ones in which profound changes in spatial domain occur and are typically misunderstood as mere national, racial, political, or cultural events. These events must be given their proper geobiological interpretation. The geography of the future must create spaces in which mankind is preferred to machinekind. Specifically, geographers must help to construct regions in which children prosper, since the protection of the species' young is the protection of the species itself.

John Fraser Hart, 1982, p. 8. The attempt to transform geography into a "science of regions" was based on the notion that regions, rather than the visible landscape, would be ideal objects of study for geographers to claim as peculiarly their own. This brand of "scientific geography" was predicated on two assumptions: first, that regions are real things, "objective realities" that are simply waiting to be discovered if only we search for them with enough diligence and intelligence; and second, that the world is covered with a network of identifiable areas, or "true geographic regions," that are homogeneous with respect to all significant geographic variables. These regions, when identified, would be valid for all purposes and accepted by all geographers. The failure of geographers to identify them was taken merely as a sign of the inadequacies of geographers, and not as a hint that such regions just might not exist at all.

John Fraser Hart, ibid., p. 9. In 1947 the A.A.G. appointed a number of distinguished geographers to a Committee on Regional Study under the kind and gentle chairmanship of Derwent Whittlesey, who seems to have been more concerned with consensus than with clarity. This Committee, after five years of deliberation, produced an elaborate and singularly fuzzy report that was so hedged with qualifications as to be more confusing than helpful. For example, saying that "areas homogeneous in terms of single features are to be considered as regions only when they are shown to possess areal qualities accordant with one or more other regional systems" comes parlously close to saying that a region is not a region unless it is a region. The report reminds one of the old saw that a camel is a horse designed by a committee.

John Fraser Hart, ibid., p. 15. An approach that has difficulty dealing with whole people and whole places leaves some geographers with a vague feeling of uneasiness, because it misses the complex reality of people and place that makes geography so fascinating for them.

John Fraser Hart, ibid., p. 21. The region in geography is analogous to the era in history. It is a pedagogic device for organizing and explaining a complex world, and its usefulness is directly proportional to its ability to facilitate an understanding and an appreciation of that complexity.

Richard Morrill, 1983, p. 6. Regions may be defined as the territories which result from the playing out of physical and human processes in the landscape. They are the manifest taxonomy of geography, the empirical, interrelated composites of phenomena which geographers, as both scientists and humanists, strive to explain and understand.

John R. Borchert, 1983, p. 147. Epochs like regions are absolutely necessary, but insufficient generalizations to describe a turbulent stream of historical geographical change.

Allan Pred, 1984, p. 279. Settled places and regions, however arbitrarily defined, are the essence of human geographic inquiry.

Alan L. Laity, 1984, p. 290. A region can . . . be defined as the union of specified elements which exist independently of thought, and are considered to be associated for the purposes of the observer. Therefore, all objects are regions, and all regions are objects.

## AREAL DIFFERENTIATION

F. J. Turner, 1914, p. 593. In Illinois a map of party grouping looks like a map of the original forest and prairie areas, with the glacial lobe extending from Lake Michigan clearly visible. In eastern Wisconsin and in the Illinois [and Missouri] counties adjacent to St. Louis, the German area emerges, in the former as a group of Democratic counties in a Republican region . . . and in the latter regularly as a group of . . . Republican counties in a Democratic area.

R. H. Whitbeck, 1914, p. 413. Two hundred years ago Nova Scotia was nearly if not quite as promising a colony as Massachusetts. Its position was more strategic. Its climate was as good and its resources were superior. Massachusetts has no equal area of farm land as fertile as the Annapolis Valley.

Mark Jefferson, 1915, p. 140. ("Regional Characters.") The details in the growth of our cities reveal an intimate relation to their environment. Thirty-four cities of the humid East show *vigorous* growth, adding larger numbers to their population with each decade. The river cities of the humid East after early vigor show *halting* growth, as the railroads superseded river transportation. The Great Lakes cities show symptoms of the same disease since 1880. When the transcontinental railways opened the Pacific Coast to the American people, who were adventurous and well equipped with material resources, there resulted an *exuberant* growth of the cities of the new West, doubling their population for two decades running. Washington, though earlier vigorous, shows distinct signs of atrophy since its disfranchisement.

William M. Davis, 1922, p. 28. The study of home geography does not find its chief recommendation in the local information that it provides, but rather in the aid that it furnishes thru local examples to the general study of geography, by giving full meaning and reality to geographical facts and relationships the world over. The reason for this is that geography as a whole is hardly more than a combination of innumerable local or home geographies. However the home geographies of different places may vary, the distant ones can always be better appreciated if the local one is consciously observed and understood as a member of the class to which it belongs.

Carl O. Sauer, 1924, p. 17. The general tendency is to consider geography in its central themes as a social science, concerned with the areal differentiations of population and occupation. In this view areal relation of human groups becomes the identifying mark of geographic studies, as time relation does of history. Distributional differences of life therefore represent the distinctive field of geography. The first task is to chart such significant distributions; the next is to explain them.

Carl O. Sauer, ibid., p. 32. All about us lies a great and essentially uncultivated field of geography. The strange and distant scene has born an unholy charm to the geographer who has thought that travel in far lands is the beginning of geographic research. The "grand tour" has long been recognized as part of the training of the man of culture. But what special competence have we to evaluate the geography of a country, whereof we understand but poorly the language and institutions, except as we have prepared ourselves carefully for comparative study by an immersion into the problems of familiar areas? Then only can we discover truly the significant contrasts of far countries.

Clarence F. Jones, 1927, p. 139. Chile, the smallest of the A-B-C Republics, holds third place in the commerce of South America, with 10 percent of the trade of the continent. The value of its commerce amounts to more than one-half that of Brazil, whose vast extent embraces 45 percent of the area of South America, and about one-fifth of that of Argentina, the premier commercial republic of the southern hemisphere.

George G. Chisholm, 1927, p. 287. I do not believe that the different parts of the world can be completely assimilated to one another. But there are intimate reactions between even remote parts of the world, with tendencies to more and more rapid response. Such reactions took place even in the distant past and, in some cases, between peoples who at the extreme limits of reaction knew nothing of each other. Such, for instance, seems to have been the case with the silk trade carried on in ancient times between China and the countries around the Mediterranean, when in China there was probably nothing known of the countries ultimately reached by the silks, and in these countries nothing of the country of origin. Probably, too, there was only comparatively little known of each other by the countries that carried on the much more important spice trade which had so great an influence down to the end of the fifteenth century on the fortunes of many great cities on the Mediterranean and between it and India.

Francis Delaisi, 1927, pp. 138–140. M. Durand begins his day by washing himself with soap made from Congo nuts, and drying himself with a towel made in Lancashire from cotton grown in Texas. He puts on a shirt and collar of Russian linen, a suit made of wool from Australia or the Cape, a silk tie woven from Japanese cocoons, and shoes whose leather came from an Argentine ox, and has been tanned with German chemicals. In his dining-room, which is adorned with a Dutch sideboard made of wood from Hungarian forests, the table is furnished with spoons and forks of plated metal, made of Rio Tinto copper, tin from the Straits and Australian silver. His fresh loaf is of wheat from the Beauce, Rumania or Canada, according to the season; he has a slice of chilled lamb from the Argentine, with tinned peas from California; his sweet includes English jam made of French fruit and Cuban sugar, and his excellent coffee comes from Brazil. He goes to his office in an American car; and after noting the quotations of the Liverpool, London, Amsterdam and Yokohama exchanges, he dictates his correspondence, which is taken down on an English typewriter, and signed with an American fountain-pen. In his factory fancy goods for Brazilian customers are being manufactured of materials of many origins, by machinery made in Lorraine on a German patent, and fed with English coal; he gives instructions that they are to be sent to Rio by the first German boat sailing from Cherbourg. He then goes to his bank, to pay in a cheque in guilders from a Dutch client, and to buy sterling to pay for English goods. After a profitable day, he proposes to spend the evening at a show with his wife. She dons her blue fox fur from Siberia, and her diamonds from the Cape; they dine in an Italian restaurant, go to see the Russian ballet, and after supper at a Caucasian cabaret to the music of a negro jazz band, they return home. As M. Durand falls asleep under his quilt of Norwegian duck-feathers, he thinks with pride of the greatness of France, entirely self-supporting and able to snap her fingers at the whole world.

G. T. Renner, 1930, p. 349. The study of geography should logically begin with the homeland and the local region. By doing so, the child begins with the tangible actualities of the subject and progresses outward from the known to the unknown.

Loyal Durand, Jr., 1932, p. 55. There are twenty-two geographic regions in Wisconsin.

John K. Wright, 1932, p. 354. In many respects our history reflects our European origins. Yet we are not European. Nearly everything that has come from the Old World has undergone a subtle but profound transformation amid a new environment. This, perhaps, is why political and social movements in the United States often present an enigma to the European observer.

V. C. Finch, 1934, p. 119. Having made alive the landscape of the whole area the geographer is next faced with the task of subdividing it into component localities upon a satisfactory basis.

Isaiah Bowman, 1934, p. 119. (''Geography in Relation.'') Geography takes the earth region by region and as a whole and works out such conclusions as its data warrant: as in all human studies the descriptive element is large, the explanatory element small.

John Kerr Rose, 1935, p. 53. Areal differentiation is a major objective of geographical study.

George B. Cressey, 1935, p. 297. Is Eurasia one continent or two?

John K. Wright, 1936, p. 620. Countless pictures and descriptions have made all the world familiar with the stupendous sky line of southern Manhattan's skyscrapers. The island's opposite end, some fourteen miles to the north, is totally different in aspect. There Manhattan terminates at the foot of a pleasant wooded hill, unmarred until recently by man-made structures and suggestive even today of the wilderness that once covered the whole region within the present city limits.

Richard Hartshorne, 1937, p. 162. The geographic confusion of ethnographic, economic, and strategic relationships which constitute the ''Corridor problem'' has existed in one form or another since the Middle Ages. In general it is but one of several problems arising from the absence in the north European plain of any strong natural divides separating the settlements of German and Slavic peoples.

Richard Hartshorne, 1938, p. 276. One of the most difficult problems in the political geography of European countries, that of national minorities, is happily lacking in the United States. We have no areas in which the majority of the residents are both culturally and nationally associated with outside countries rather than with the United States.

Richard Hartshorne, 1939, p. 305. The problem of determining and explaining the where of things is not the distinctive function of geography.

Richard Hartshorne, ibid., p. 307. Geography, in brief, can demand serious attention if it strives to provide complete, accurate, and organized knowledge to satisfy man's curiosity about how things differ in the different parts of the world, just as history in similar fashion strives to satisfy his curiosity about what things were like in the past; and just as history considers the past in terms of periods because men live and things happen together only within a limited space of time, so geography must consider the world in terms of limited areas within which things are closely associated.

Richard Hartshorne, ibid., p. 548. Geography does not claim any particular phenomena as distinctly its own, but rather studies all phenomena that are significantly integrated in the areas which it studies, regardless of the fact that those phenomena may be of concern to other students from a different point of view. . . . Geography need not look for any concrete objects as its own.

Richard Hartshorne, ibid., p. 636. In its historical development geography has occupied a logically defensible position among the sciences as one of the chorographical studies, which, like the historical studies, attempt to consider not particular kinds of objects and phenomena in reality but actual sections of reality; which attempt to analyze and synthesize not processes of phenomena, but the associations of phenomena as related in sections of reality.

Richard Hartshorne, ibid. Whereas the historical studies consider temporal sections of reality, the chorographical studies consider spatial sections; geography, in particular, studies the spatial sections of the earth's surface, of the world. Geography is therefore true to its name; it studies the world, seeking to describe, and to interpret, the differences among its different parts, as seen at any one time, commonly the present time. This field it shares with no other branch of science; rather it brings together in this field parts of many other sciences. These parts, however, it does not merely add together in some convenient organization. The heterogeneous phenomena which these other sciences study by classes are not merely mixed together in terms of physical juxtaposition in the earth surface, but are causally interrelated in complex areal combinations. Geography must integrate the materials that other sciences study separately, in terms of the actual integrations which the heterogeneous phenomena form in different parts of the world.

Richard Hartshorne, ibid., p. 639. Geography seeks to acquire a complete knowledge of the areal differentiation of the world, and therefore discriminates among the phenomena that vary in different parts of the world only in terms of their geographic significance—i.e., their relation to the total differentiation of areas.

Richard Hartshorne, ibid., p. 641. Geography seeks to organize its knowledge of the world into interconnected systems, in order that any particular fragment of knowledge may be related to all others that bear upon it. The areal differentiation of the world involves the integration, for all points of the earth's surface, of the resultant of many interrelated, but in part independent, variables. The simultaneous integration all over the world of the resultant of all these variables cannot be organized into a single system.

Richard Hartshorne, ibid. In systematic geography each particular element, or element-complex, that is geographically significant, is studied in terms of its relation to the total differentiation of areas, as it varies from place to place over the world, or any part of it.

Robert M. Glendinning, 1940, p. 56. The dry Death Valley trench stands in sharp contrast to the humid and irrigated valleys of California.

Carl O. Sauer, 1941, p. 2. ("Foreword.") A peculiarity of our American geographical tradition has been its lack of interest in historical processes and sequences, even the outright rejection thereof. A second peculiarity of American geography has been the attempt to slough off to other disciplines the fields of physical geography. Hartshorne's recent methodologic study is an interesting illustration of both these attitudes. In spite of basing himself strongly on Hettner, he does not take into account the fact that Hettner's own contributions to knowledge have been chiefly in the field of physical geography. Nor does he follow Hettner into his main methodologic position, namely that geography, in any of its branches, must be a genetic science, that is, account for origins and processes. Hartshorne, however, directs his dialectics against historical geography, giving it tolerance only at the outer fringes of the subject. I have cited this position because it is the latest and, I think best statement of what is in fact, if not by avowal, a pretty general viewpoint in this country.

Carl O. Sauer, 1941, p. 353. ("Personality.") This is an excursion into the oldest tradition of geography. For whatever the problems of the day may be which claim the attention of the specialist and which result in more precise methods of inspection and more formal systems of comparison, there remains a form of geographic curiosity that is never contained by systems. It is the art of seeing how land and life have come to differ from one part of the earth to another. This quality of understanding has interested men almost from the beginning of human time and requires restatement and reëxamination for each new generation.

Richard Hartshorne, 1948, p. 121. If it be agreed that it is no criticism of a study made by a geographer to say that it is not geography, is it not a criticism of the geographer who made this study? This common attitude is based on assumption that those who call themselves geographers should be working at geography. Regardless of the significance or value of a particular study, if it cannot be classified as geography, the geographer should not have made the study.

Robert S. Platt, 1948, p. 132. ("Determinism.") Thus geography deals with perceived similarities and differences between areas, perceived in the first instance rather vaguely and in the end given a more precise definition, quantitative in some cases, tracing covariation in space, and thus approaching "explanation in the only scientific way, that of observed coincidence."

Stephen S. Visher, 1952, p. 98. The birthplaces of the numerous popular authors and journalists, distinguished scientists, and prominent politicians born in Indiana are concentrated much more than the population was at the time of their birth. A fifth of the state, then having about one-fourth of the population, was the birthplace of two-thirds or more of the prominent people born before 1970. Conversely, an area of even greater population produced very few of those notables. The productive area was to the east and south-east of Indianapolis; the most unproductive one was in the north-western part of the state.

Glenn T. Trewartha, 1953, p. 83. In our science, the central theme of which is areal differentiation, the dynamic and pivotal element is human life, or population.

Carl O. Sauer, 1956, p. 297. The conventional areal study may be an encyclopedia but it is not a synthesis. Are we not under a form of inductive fallacy if we collect lots of data on lots of subjects thinking these will somehow acquire meaning?

Richard Hartshorne, 1958, p. 108. ("Concept of Geography.") The intrinsic characteristics of geography are the product of man's effort to know and understand the combinations of phenomena as they exist in areal interrelation in his world.

Richard Hartshorne, ibid. In particular, geographers from early times have observed that work in their field differs from that in many other sciences in the following respects: (1) the fact that geography has no one particular category of objects or phenomena as its specific subject of study but studies a multitude of heterogeneous things as integrated in areas; (2) geography cannot be classified as either a natural science or a social science, nor simply as a bridge between the two groups, but rather must study combinations in which both kinds of phenomena are intimately intermixed; (3) study in geography requires the use of two markedly different methods of study: the systematic examination of certain categories of relationships over the world or any large part of it, in general or systematic geography; and the study of the totality of interrelated phenomena in particular areas, in special or regional geography; and (4) while geography like all other sciences is concerned with the development and application of generic concepts and general principles or scientific laws, it is like history in that it is also concerned in large degree with the knowledge and understanding of individual, unique areas.

Wilbur Zelinsky, 1958, p. 229. Although it is clear that we can never aspire to simple formulae for defining, gauging, and abating "underdevelopment," it is equally obvious that a firm grasp of the population factor is indispensable for any real progress towards understanding the problem.

J. E. Spencer, 1958, p. 290. The term "underdeveloped" is now popular in reference to particular portions of the world lacking certain attributes. Unless "underdeveloped" is defined by a single criterion, there must be many kinds of "underdevelopedness." If an area is to be termed "underdeveloped," is the reference to things, to conditions, or to people? One aspect of many generalizations on the "underdeveloped" world has been reference to too many people. How many is too many, and how many is too few?

F. Lukerman, 1962, p. 348. A place exists, has configuration, size, shape, position, and can be so described, but it is much more. A place is an association of ideas—impressions—held by men and believed by men to reside in a specific complex of integrated phenomena. A place is a view, an image, and geography, as the study of places making up man's habitat, is a view of the world, an imago mundi.

Richard Hartshorne, 1962, p. 10. For more than 30 years the phrase "areal differentiation" was accepted by many American geographers as a technical term expressing most briefly our view of the subject matter of our field. When however I came, in 1959, to examine empirically the varied meanings which different geographers found in the phrase, I was forced to conclude that it was inadequate to provide an explicit, clear expression of the concept it was presumed to identify, and therefore, if used as a convenient shorthand label, it should be used only among professional colleagues who have learned what it is intended to represent.

J. M. Blaut, 1962, p. 7. Areal integrations, viewed as the distinctive subject matter of geography, provide us with that special something which we all seek but some despair of finding: a simple, clear definition of our field; a basic organizing concept.

Barry Floyd, 1963, p. 15. Geography's *raison d'être* depends upon the ability of its practitioners to continue mastering and integrating knowledge from many other fields in their pursuit of wisdom with respect to place and their quest for the "living tether."

Edward A. Ackerman, 1963, pp. 431–432. [Areal differentiation] favored . . . a goal of investigation independent of the goals of other sciences.

Karel J. Kansky, 1964, p. 426. It may be observed that the structure or layout of transportation networks varies from region to region.

Norton Ginsburg, 1973, p. 15. One objective of a locational strategy in development would be to lessen the historical contradictions between "urbanism" and "ruralism" as ways of life.

Edward J. Taaffe, 1974, p. 6. On the positive side, the integrative view showed us that widely disparate phenomena may be closely interrelated. The artificial separation of phenomena into compartments marked economic, political or social often appears to be more of a curricular convenience than a replication of reality or an operationally useful way of dealing with problems.

Edward J. Taaffe, ibid. Another weakness of the integrated view at this time lay in what should have been its greatest strength. The very thing it should have done most effectively, namely, to bring geographers into closer contact with the work and ideas of other social scientists, it failed to do.

Marvin W. Mikesell, 1978, p. 4. Migration, diffusion, expansion, transfer—these words identify the orientation of cultural geographers even more explicitly than the theme of change.

Marvin W. Mikesell, ibid. Among the preferences that define the personality of cultural geography mention should be made of (1) a historical orientation, (2) stress of man's role as an agent of environmental modification, (3) a preoccupation with material culture, (4) a bias in favor of rural areas in this country and non-Western or preindustrial societies abroad, (5) a tendency to seek support from anthropology, (6) a commitment to substantive research and a consequent attitude of extreme individualism, and (7) a preference for field work rather than "armchair geography." Although these preferences are not universal among cultural geographers, they are sufficiently common, I think, to provide a foundation for understanding of their behavior.

Marvin W. Mikesell, ibid., p. 14. Understanding of what has been common in the causes of Catholics in Ulster, Muslims in Lebanon, Bengalis in Pakistan, or Ibos in Nigeria requires more than geographical skills. Yet the distinction between "we" and "they" or between "have nots" and "haves" calls for delimitation as well as redress, and delimitation is the one effort we can claim to accomplish more effectively than any one else.

John Fraser Hart, 1982, p. 23. Regional geographers are concerned primarily with patterns and the interrelationships of patterns, and they are concerned with processes only as they are necessary to understand patterns. Regional geographers must be especially sensitive to the distinction between geographical correspondences and causal relationships, and they should not impute cause unless they can demonstrate the responsible process. They assign cause, however, because inferential statisticians labor under precisely the same constraints.

John Fraser Hart, ibid., p. 27. One of the more serious weaknesses of contemporary American geographers may be their lack of field experience in foreign areas, because they cannot understand the distinctiveness of their own part of the world until they know how it differs from other parts.

Hendrik-Jan A. Reitsma, 1982, p. 125. Since the 1950's, it has become more or less common practice to divide the world into three broad, noncontiguous regions: The First World (i.e., the capitalist developed countries), the Second World (all socialist states), and the Third World (the nonsocialist underdeveloped countries), . . .

It should be noted, however, that this happens to be a most peculiar classification, not unlike one that divides mankind into rich, poor, and tall people.

## REGIONAL CHANGE

Mark Jefferson, 1914, pp. 401–402. All countries of the world that have a regular census show an unfailing growth of population—even through great wars like our Civil War and the Franco-German War of 1870—excepting only India between 1861 and 1871 and Ireland, if Ireland can be called a country.

Mark Jefferson, 1915, pp. 20–21. How effectively the Appalachians bar other eastern cities from easy intercourse with the continental hinterland we see by the growth of Boston, of Philadelphia, and, most of all, of Baltimore. Cities on our inland waters flourished with the early development of river and lake navigation, only to suffer an inevitable check as the hand of railroad control choked out the life of water-carriage, as happened to Cincinnati forty years ago and to Chicago during the last two decades in much less degree.

Preston E. James, 1929, p. 67. To the geographer, Southern New England offers a fascinating and challenging problem in landscape interpretation. Occupied successively by the Indians, by European agriculturalists and by manufacturers, the several cultures have left their own peculiar impressions in the landscape. Thus a confusing array of earlier and later forms, of impression set on impression is presented to the eye. Southern New England, with its long history and its several periods of development, possesses a landscape composed of a complex of relict forms, intricately involved with the more recent forms.

W. Elmer Ekblaw, 1931, p. 121. Russia today is essentially the Russia of yesterday and the day before despite the change from Czarism to Communism.

Derwent Whittlesey, 1931, p. 143. In the chorologic landscape can be read the sequence and the character of terrene occupance. A dilapidated business street fronts the decayed wharves. Its crossing of Main Street established the business center. Astride its prolongation toward the falls stands the civic center on high ground. The residential sections, chiefly of frame, story-and-a-half cottages and simple two-story houses, cling to the slopes bordering the estuary and parallel the river to the uppermost fall. On every country road beyond the uninterruptedly built-up area are shacks of the poorest villagers.

W. M. Davis, 1934, p. 11. (''Long Beach.'') Life is full of hazards, and we must take our chances among them. The chances of an enjoyable life in southern California are, in spite of its occasional earthquakes, undeniably excellent. [Written in the year he died, at the age of 84.]

Mark Jefferson, 1934, p. 634. In the United States children under three years of age were fewer in 1930 than in 1920. Presumably the country is somewhere near that transition stage of population growth in which it is difficult to observe the moment when growth gives place to decline.

G. Donald Hudson, 1935, p. 275. These then are four major contributions that geography can make to regional planning: (1) geography's broad regional perspective; (2) sound thinking with regard to the relationships between human activities and their physical settings; (3) geography's scientific approach; and (4) methods and techniques of geographic investigation and presentation.

Derwent Whittlesey, 1937, p. 170. The only truly reliable source of geographic information is the visible landscape of the present.

Jean Gottmann, 1937, p. 550. The last twelve years have witnessed a rapid economic development in Palestine through the men and money brought by Jewish colonization. Unhappily this development has been accompanied by political strife, even by sanguinary disorders. The task assumed by Jewish colonization—to gather together on the land of their ancestors a people dispersed two millenniums ago to the four corners of the earth—is an enterprise without precedent that cannot fail to create complex political problems.

Julie Moscheles, 1937, p. 414. Urban geography is mostly concerned with the geographical interpretation of the origin, character, and development of towns.

J. Russell Smith, 1939, pp. 94–95. Over-pasturing on the uplands of New Mexico is causing increased run-off and increased erosion by the streams flowing into the Rio Grande in that state. This is a perfect example of regional suicide.

Carl O. Sauer, 1941, p. 8. (''Foreword.'') The whole task of human geography, therefore, is nothing less than comparative study of areally localized cultures, whether or not we call the descriptive content the cultural landscape. But culture is a learned and conventionalized activity of a group that occupies an area. A culture trait or complex originates at a certain time in a particular locality. It gains acceptance, that is, is learned by a group, and is communicated outward, or diffuses until it encounters sufficient resistance, as from unsuitable physical conditions, from alternative traits, or from disparity of cultural level. These are processes involving time and not simply chronologic time, but especially those moments of culture history when the group possesses the energy of invention or the receptivity to acquire new ways.

Loyal Durand, Jr., 1941, p. 289. Cities, localities, and regions frequently obtain reputations for their products—reputations which become valuable advertising assets beyond the home community, and reputations which may in turn attract other manufacturers or dispensers of the noted product to the city, locality, or region. States, likewise, attain characteristic industries and attitudes, and a "feeling" may come to pervade the inhabitants of the area—a pride in the enterprise. The "state characteristic," difficult to measure objectively, sweeps from border to border, oftentimes stamping the landscape with undoubted elements of importance, and carrying the industry or enterprise into regions not always the most ideally suited to it.

Alberto Area Parró, 1942, p. 1. Peru, after 64 years, has taken a population census. It is hard to understand how she could have got along without one for so long a time. What is the explanation? What happened to the statistical experience of Inca and colonial times?

Chauncy D. Harris, 1942, p. 313. Cities in the United States had shown consistent and rapid growth for many decades before 1930, but the Census of 1940 revealed that during the decade 1930–1940 many cities actually lost population.

Walter W. Ristow, 1944, p. 333. The implications of present and future air routes over hitherto barren and inaccessible lands are comparable, and, in some ways, even more revolutionary, than the changes in geographical thinking which resulted from the discoveries of Columbus or the invention of the steamboat. All geography becomes *home* geography when the most distant point on the earth is less than sixty hours from your local airport. Air Age geography thus is *world* geography, with the world greatly reduced in size as measured by the fourth dimension of *time*. All nations are truly neighbors, and now more than ever before, they must learn to live in peace and harmony.

Raymond E. Crist, 1955, p. 176. By revisiting the area in 1954, I was able to get historical perspective on changes in transportation, farming, public health, and so on, in the area, and analyze the reasons for such change.

J. Wreford Watson, 1958, p. 381. In a changing world few continents earn more jealousy yet excite less envy than North America. It earns the jealousy of all who fail to adapt themselves to change as rapidly or successfully. The one great advantage of America is that it meets change with change. It has the most dynamic and productive way of life on earth. America's today is the world's tomorrow. And yet it invites little envy; increasing power compels it to carry increasing responsibility. Heading means leading; to acquire assets is to accept liabilities.

Andrew H. Clark, 1962, p. 235. In no better way can a region be brought to life than by recreating its course and character of past change.

William Bunge, 1964, p. 28. The South of the United States has long been an area of social and economic backwardness and is now the center of a revolution that could conceivably not only bring the region to national standards but beyond these improvable standards. Where the worst previously existed, the best might well flower.

Fred Kniffen, 1965, p. 549. In America, settlement geography has somehow failed to find equally widespread acceptance. A major reason, I believe, lies in the brevity of our historical time span. The methods of historical reconstruction employed in Europe are less applicable here.

Fred Kniffen, ibid. It is still possible in the eastern United States to distinguish the initial occupance patterns established by migrants from the seaboard areas. Among the occupance ingredients, housing of all kinds is the most reliably diagnostic of the whole; however, the course of development from each of the source areas followed quite different lines.

John Fraser Hart, 1968, p. 417. The frontier tradition of taming the wilderness, clearing the forest, and bringing the land into agricultural production is so deeply engrained in the national mythos of the United States that many Americans refuse even to consider the possibility that land won with such difficulty could ever be abandoned; the very notion seems sinful. The most common reaction, when confronted with data indicating that the acreage of farm land is declining, is to assume that the former farm land has been converted to urban uses. This is considered a sign of progress, whereas the idea of abandonment has connotations of retrogression.

John R. Borchert, 1971, p. 22. That heterogeneous procession and many other geographers have passed along the trail through the Prairies and Plains. They have studied the settlements and the resources of this dynamic testing ground that tapers eastward from the American cordillera in the Northern Hemisphere Westerlies and widens westward from the American Manufacturing Belt in the Northern Hemisphere Transportation Network. They have combined concept and technique with the specifics of a regional problem to contribute to greater understanding of how human settlement evolves as one of man's technologies to utilize the earth. Hence they have helped Geography to contribute to the Grand Probe of Science into the nature, the purpose, the goals, and the destiny of mankind.

Today this region, among many, cries out for more precise analysis in terms of interacting systems. As one looks ahead into the 1970s, it seems likely that growing numbers of geographers will pack their improved tools from the preceding Decade of Progress and head once more for the Field, to continue the Grand Probe of Science where countless regions such as this one lie richly carpeted with unworked data and unsolved problems.

Richard V. Francaviglia, 1971, p. 509. Cemeteries are deliberately created and highly organized cultural landscapes.

Norton Ginsburg, 1973, p. 5. Around the turn of this century or shortly thereafter, the world probably will continue to be divisible into two major parts—a developed world and an underdeveloped one; that is, some nations that are comparatively rich, and many that are poor. Most of the six billion people in the world at that time will live in the latter type of country. They will reflect the wisdom of Jesus that, "the poor always ye shall have with you"; and the numbers of the poor will be increasing considerably more rapidly than the rich.

Norton Ginsburg, ibid., p. 6. Since science has been concerned primarily with problems relating to the richer countries, and these are mostly Western as well as middle-latitude, its concepts are ill-suited to the appropriate definition of problems in other contexts, let alone to the development of technologies suited to solving them.

Norton Ginsburg, ibid., p. 13. By the turn of the century it is probable that well over half of the urban dwellers in the world will live in cities in Asia, although the proportions of population classified urban in Asian countries will continue to be low, with few exceptions.

Marvin W. Mikesell, 1978, p. 6. While economic geographers have been preoccupied with processes that result in centralization and standardization, cultural geographers have been engaged in studies of communities that have managed to retain essential features or at least some features of their historical personalities.

John Fraser Hart, 1982, p. 24. Traditional geography has failed to realize that each city has its own regional geography. The new frontiers of regional geography may lie more in our great cities than in the countryside.

John Fraser Hart, ibid., p. 27. The study of a region must be such a lifetime commitment that a geographer should choose his or her region with great care. You should select a region in which you are so interested that you are willing to dedicate a major part of your life to getting to know it intimately.

## REGIONAL GEOGRAPHY

Charles Redway Dryer, 1912, p. 73. Regional geography is undoubtedly "the bright, consummate flower" of geographic science, from which the best fruit of scientific method and discipline is to be obtained.

W. M. Davis, 1915, p. 62. True regional geography is the final object of a geographer's efforts.

Wellington D. Jones and Carl O. Sauer, 1915, p. 521. If the region is unfamiliar, it is usually best to begin with a rapid trip through the area, to get general relations in mind.

Nevin M. Fenneman, 1919, p. 7. The one thing that is first, last, and always geography and nothing else, is the study of areas in their compositeness or complexity, that is *regional geography*.

Charles Redway Dryer, 1920, p. 11 It is in the regional form that geography has finally attained consciousness of its object, its methods, its function,—of itself.

Albert Perry Brigham, 1920, p. 100. Geography is to be a part of the training of statesmen. One of the reasons, perhaps the great reason, for thrusting geography, a generation ago, into the classic conservatism of Oxford and Cambridge, was that some gentlemen in London thought it was absurd for British public men to rule in geographic ignorance an empire that followed the sun.

Albert Perry Brigham, ibid. We must therefore have geographic training for citizenship. Putting the people in the saddle is very well if the people are fit to ride and know where they are going. The ungeographic voter is pretty sure to derive his opinion from a blatant neighbor, or a hidebound editorial writer, or from the party boss in his neighborhood.

Charles Redway Dryer, 1920, p. 3. The late war has taught the world, among other things, that geography is one of the sciences which have a practical bearing upon the largest affairs of the world. Perhaps for the first time in history, geographers have been summoned to the councils of the nation to perform duties which only men with geographical training can perform.

Carl O. Sauer, 1921, p. 3. ("Land Classification.") In regional geography there are two leading questions to be answered: To what uses is the area in question being put; and: What are its possibilities? On these two questions the fundamental technique of regional analysis rests.

C. E. Marston, 1921, p. 311. The memory "crazy quilt" type of geography teaching is still common in our schools and accounts for the general dislike and lack of interest pupils manifest in the subject. It is not to be inferred that vital facts are to be neglected. They will be presented as explanatory of other facts and will be grounded naturally in support of broad generalizations. Even more facts than now taught can be retained if properly associated in an interesting way. We have followed a faulty psychology in drilling pupils on unrelated geographic material.

William M. Davis, 1922, p. 28. There are certain important principles that the teacher should bear in mind during the progress of local study. Geography teaches us about the way that people live on the earth—this being a rough conversational definition of this subject, sufficing to embody in elementary form "the study of the study of the earth in relation to man." Geography is therefore concerned with two classes of facts and with the relations in which the two classes stand. The first class embraces all necessary facts about the inorganic earth—land, water, air—and about plants and animals considered as the non-human inhabitants of the earth; the second includes the necessary facts as to the manner of man's living, from the savage to the civilized state, from wandering nomads to fixed populations, from the thieving of warfare to the competition of trade. It is only as the facts that constitute these two classes come to be understood that their relationship can be studied; and this matter of relationship then becomes the very soul of geography.

Harlan H. Barrows, 1923, p. 10. Urban geography really is a phase of regional geography.

E. E. Lackey, 1924, p. 68. Now, home geography presents several difficulties. There are elements that are suitable for presentation to the earliest grades, other phases adapted to the upper grades, and some portions of sufficient difficulty to engage the best efforts of high school and college students. Home geography is a regional study whether presented in the elementary school or later.

W. M. Davis, 1924, p. 210. Any person who knows a region so well by long residence in it and by much travel over it that he can give an accurate account of its topography, of its climate and weather, of its soils especially as shown in its fields when plowed and as reflected in the crops that the fields bear, of its natural plant cover and its cultivated crops varying in appearance with the seasons, of its animals as they appear in the landscape, of its human inhabitants with their works and customs and activities in so far as they are related to the region—the account being so constructed as to treat each of these factors reasonably, to embody the correlations of all the factors one with another as they exist together, and to present the total picture intelligibly to a competent reader—such a person may justly call himself a geographer.

Frederick Jackson Turner, 1926, p. 35. As the United States becomes a settled nation, about equal to Europe in area and resources, with provinces, or sections, analogous to European nations, regional geography demands increasing attention.

Glenn T. Trewartha, 1932, pp. 80–81. The author contends that a morphological study of any area requires no defense since it deals with a portion of the earth's surface.

V. C. Finch, 1934, pp. 113–114. The structures of papers in scientific regional geography should be so articulated that the regions they present are capable of comparative analysis. To the unlearned soldier, familiar as he may be with dead men's bones and the sight of decaying horse-flesh, the heel of a man and the hock of a horse have no significant connection. There is in his mind no structural basis of comparison that serves to bridge the gap between the individual and the group, or to turn confusion into comprehension. Only through a study of the comparison of geographic forms, patterns and associations shall we arrive at anything approaching a scientific regional geography of the world. However, no effective comparison is possible unless the papers that are written in the field of regional geography are cast in molds whose forms are known and whose parts are capable of being arranged in an understandable order so that the items of studies cast in the same mold or even in different molds may be set beside one another and the degree of their accordance or discordance discovered.

Wellington D. Jones, 1934, p. 105. Most geographers agree that one of the major objectives in regional investigation is the determination of areas of various magnitudes that display significant homogeneity.

A. E. Parkins, 1934, p. 225. Geography is a systematic study of those facts in a region or an area which can be studied distributively, and of the relation existing between these groups of facts, either human items or environmental.

A. E. Parkins, ibid. Geography is the science of areas.

Robert Burnett Hall, 1935, p. 123. Geography is a synoptic science and as such must take cognizance of the problems and advantages which the region presents to the other sciences involved. In fact, geography must by its very nature as the space science, face all of the problems and receive all of the benefits which the region has to offer.

Robert Burnett Hall, ibid. In geography, the region forms the basis of what is now termed modern or regional geography and has come to be widely accepted as the "culminating branch of the science." To many workers in the field, regional geography is geography.

Robert Burnett Hall, ibid., p. 124. The development of the regional concept in geography has proceeded through a number of distinct stages, each more ambitious than the preceding one and roughly correlated with the expansion of accumulated knowledge about the lands of the world.

Robert Burnett Hall, ibid., p. 126. The approved technique of modern regional geography might be stated as follows, first, the mapping of the critical forms and patterns as to distribution, secondly, the assemblage of these as to their genetic relationships, thirdly, ascertaining when possible their origins and developments, and finally, attempting to synthesize them into a regional pattern. The wide and rapid acceptance of this latest concept of the geographical region was accompanied by a wave of intense optimism. By some the new terminology was swallowed whole but the new ideas seem not to have been understood and the region came to serve as a cloak for further studies in determinism. Others thought they saw in it an opportunity to make geography an exact science and, in their haste, created false truths by mathematical formulae. The great majority, however, worked hard and intelligently in the middle of the field—regional analysis.

V. C. Finch, 1939, p. 11. Many geographers have pondered over the nature of regions.

Richard Hartshorne, 1939, p. 306. Geography as a chorographic (or chorologic) study has always found its justification in the widespread desire of many people to know what other parts of the world are like, just as history finds ample justification in the common desire to know what happened and what things were like in past times. Further, since the ordinary person actually knows but partially the area in which he lives, just as he understands but little of what is happening in the world in his lifetime, "home geography" is as necessary as current history.

Richard Hartshorne, ibid., p. 322. Almost all geographers agree that the clearest view of geography is to be seen in regional geography.

Richard Hartshorne, ibid., p. 612. The development of geography during the past thirty-odd years has been marked by an increased interest in regional geography.

Richard Hartshorne, ibid., p. 618. Regional geography, therefore, studies the manner in which districts are grouped and connected in larger areas, the manner in which these areas are related in areas of greater scale, and so on, until one reaches the final unit, the only real unit area, the world.

Richard Hartshorne, ibid., p. 643. Regional geography organizes the knowledge of all interrelated forms of areal differentiation in individual units of area, which it must organize into a system of division and subdivision of the total earth surface.

Edward A. Ackerman, 1945, p. 123. Regional geography and systematic geography were, and still are thought of by many American scholars as two different and almost incompatible manners of study, with different objectives and different techniques. This belief is so much a part of the folklore of our field that it is accepted by nearly all authors without discussion. In the few recent instances where critical analysis and logical discussion are provided, as in Hartshorne's "The Nature of Geography," the same conclusion is reached. Each of the two approaches is generally thought to lead to different ends, although the material brought forth by one is considered useful to the other. Since most geographers of the recent past have looked upon regional geography as the culmination of all study in this field, a bias toward regional techniques of study was a natural development.

Henry J. Warman, 1946, p. 173. If the urge to study global or world geography has accomplished anything at all we must accept the fact that it has caused all geography to pause and size up the geography pattern as a whole.

Werner J. Cahnman, 1948, p. 241. Area studies restores the methodological balance in the social sciences.

Otis P. Starkey, 1950, p. 152. The author proposes a joint inventory to be conducted by university departments to list and plot on state maps all regional studies completed for the United States. Such studies should be assembled and distributed as a bibliography. The completed work would indicate graphically neglected areas in American geography.

Ralph H. Brown, 1951, p. 199. Thomas Jefferson, while Secretary of State, viewed the North America of his time as comprising two great geographic provinces. The one province facing inland and destined later to be expanded under his own guidance, he denominated the "landboard." The outward-facing slope, the Atlantic coastland, was, of course, the seaboard. The former term, which has perhaps never been used since, serves less to demonstrate Jefferson's originality (which some have lately doubted) than to show that he applied a handy name to a distinctive regional entity, one widely recognized at the time. Others of his contemporaries referred, with less precision and geographic insight, to the "back country" and to its inhabitants as "backwoodsmen." On the other hand, the authentic word "seaboard" has survived in the common language, identifying to all generations the eastern fringe of North America.

Preston E. James, 1952, p. 195. The regional concept constitutes the core of geography. This concept holds that the face of the earth can be marked off into areas of distinctive character; and that the complex patterns and associations of phenomena in particular places possess a legible meaning as an ensemble, which, added to the meanings derived from a study of all the parts and processes separately, provides additional perspective and additional depth of understanding. This focus of attention on particular places for the purpose of seeking a more complete understanding of the face of the earth has been the continuous, unbroken theme of geographic study through the ages.

Preston E. James, ibid., p. 196. The present writer is well aware that most of what he would like to say about the regional concept has already been said, not once, but many times.

Carl O. Sauer, 1956, p. 292. I do not accept the idea that anyone can do the geography of a region, or comparative geography, when he knows less about anything he assembles than others do, as I do not accept the notion that every geographer must be concerned with regional synthesis. The ineptly named holistic doctrine leaves me unmoved; it has produced compilations where we have needed inquiries. This is no counsel of despair but rather I wish to say that geography, like history, resists any overall organization of interests, directions, or skills, yet does not lose thereby an acknowledged position of its own kind of knowledge and of valid process of discovery and organization. In a time of exceedingly great increase of knowledge and techniques we remain in measure undelimited and, I may add, unreduced to a specific discipline. This, I think, is our nature and our destiny, this our present weakness and potential strength.

Carl O. Sauer, ibid., p. 293. What benefit of training and insight comes out of regional courses? After many years I am no closer to an answer. I think we give far too many such courses, that they may be given for indifferent reasons, and that too often they contribute little to learning or to skills. More and more the concern with regional classifications and regional boundaries leaves me cold.

Carl O. Sauer, ibid., p. 294. A good regional course is largely an individual creation out of long application, involving physical discomforts and pleasures, muscular, cutaneous, and gastric, and it has been nursed on much meditation. It demands some ability and interest in physical geography and understanding of other ways of life and how they came about. A really intimate association with other cultures is needed and slow to be acquired. To me it is a study in historical geography.

Robert B. McNee, 1956, p. 394. Regionalism has found a place, however small, in nearly every discipline. In history and political science it appears as the study of sectionalism, in anthropology as the study of cultural areas, and in biology as ecology. But it is geographers who have most studied regions *per se*, i.e., the concept and method of regionalism as distinguished from limited applications.

Robert E. Nunley, 1959, p. 204. This paper pleads for an analytical approach to all regional geography and suggests one way this can be accomplished. It suggests that, by asking and answering a few basic questions concerning the distribution of population, all the facts of the physical and cultural landscape generally considered in a regional geography can be woven into a stimulating, meaningful study.

Andrew H. Clark, 1962, p. 232. A geographer without a deep and far-ranging curiosity about the world at large surely would be undeserving of his title.

Edward A. Ackerman, 1963, p. 437. The concept of a region is potentially valuable in systems study, but we should take care that the regional concepts we actually use are significant to the overriding system.

William D. Pattison, 1964, p. 213. The area-studies tradition (otherwise known as the chorographic tradition) tended to be excluded from early American professional geography.

Lutz Holzner, 1966, p. 129. American geographers generally agree that the regional concept is a fundamental part of the geographer's point of view.

F. Hung, 1966, pp. 343–344. The spread of industrialization and urbanization has radically altered the traditional man/land relationships and their inseparable unity enshrined in the landscape, which has provided a *rationale* for regional geography. . . . Regional synthesis was for a long time considered as the peak and end of geographical work. However, . . . the dialectic between man and land applied only to rural societies, and . . . industrialization has brought the gradual dissolution of the traditional, rural, regional pattern of life. . . . Thus the basic concept and the philosophy of regional geography are inapplicable to industrial-urban societies, although they still apply to the industrially underdeveloped countries of Africa, Asia and Latin America.

Preston E. James, 1967, p. 2. Geographic study, from whatever point of view and by whatever kind of method, is concerned with areas, or segments of earth-space.

Raymond E. Crist, 1969, p. 306. Large-scale extrapolation might be the means of achieving the very poor substitute for the descriptive geography and regionalism which the New Geography would like to supplant.

William W. Bunge, 1973, p. 275. ("Human Survival.") The earth itself is a man-made region.

Leonard Guelke, 1977, p. 1. Not too long ago regional geography was widely regarded as constituting the essential core of the discipline of geography. The importance of regional work was an outgrowth of the idea that the description and explanation of the varied character of the earth's surface was the primary goal of geography. In pursuing this goal geographers emphasized that they, unlike other earth and social scientists, were fundamentally interested in associations of phenomena on the earth's surface rather than the phenomena in themselves. In practice many geographers were involved in topical or systematic studies, but the value of such work to geography as a discipline was assessed in terms of its potential contribution to regional geography. The rise of the new geography with its emphasis on spatial relationships and the use of statistical methods has coincided with a decline in the importance of regional studies. Today, regional geography, far from being considered the essential focus of the discipline, is looked upon by many as a subjective, largely descriptive type of study of minor importance to the advancement of the field.

John Fraser Hart, 1982, p. 1. Society has allocated responsibility for the study of areas to geography; this responsibility is the justification for our existence as a scholarly discipline. Most people are inherently curious, and they want to know more about the world in which they live—about their home area, about what's over the hill, about what's beyond the horizon, about what's across the sea—and it is our duty, as geographers, to satisfy their curiosity.

John Fraser Hart, ibid., p. 2. Geography is a science, but it is also an art, because understanding the meaning of area cannot be reduced to a formal process. The highest form of the geographer's art is producing good regional geography—evocative descriptions that facilitate an understanding and an appreciation of places, areas, and regions.

John Fraser Hart, ibid., p. 10. Regionalism was a shibboleth, a slogan, a code word to identify or to stigmatize those who were opposed to the contemporary world; although there were some notable exceptions, many regionalists seemed to be anti-metropolis, anti-technology, anti-centralization, anti-twentieth century or, in short, anti-quarian.

John Fraser Hart, ibid. The regionalists had turned to the past in their search for order and stability during a period of rapid social change and stress. Regionalism was never a coherent movement, but a hodgepodge of people and ideas. It included nostalgic romantics who felt the urge to live off the land and among the folk in order to regain a sense of identity, and it also included utopian visionaries who hoped to restructure society by redirecting population growth from endless metropolitan sprawl to scattered garden cities protected by green belts.

John Fraser Hart, ibid., p. 16. Regional geography, which seeks to understand whole human situations rather than people abstracted from their real world contexts, finds that the idealist approach provides more congenial framework than positivism, which seeks to create laboratory situations by isolating specific parts for detailed analysis.

John Fraser Hart, ibid., p. 17. Despite the efforts of its friends to kill it and the efforts of its enemies to bury it, however, I firmly believe that regional geography is and ought to be a major component of geography, and that it deserves a major commitment from some—if not many—of us. We still need it, whether or not we want it. The presumed corpse has a lot more life in it than many geographers have realized, and if I may change my metaphor, it is high time for us to start cultivating more assiduously.

John Fraser Hart, ibid., p. 24. Unlike scholars in other disciplines, who can restrict their attention to a limited number of variables, geographers are interested in the totality of places, and we need to see how apparently disparate things fit together in the real world.

John Fraser Hart, ibid., p. 29. Good regional geography must be grounded on keen sensitivity to the needs, wishes, and values of the people who live in the region.

Helge Nelson, 1982, p. 121. [The geographer] ought to work with the sharp inductive method of the natural scientist, he ought to seek laws in the wonderful complex of nature and human work that geography embraces. But he must not forget that as soon as he has found laws, he should again search for the particular and the *individual* in his area of study in order to reach a synthesis and a *view of the whole* of the nonliving and living nature, of nature and culture in their settled relations.

Richard Morrill, 1983, p. 7. One major role of geography is the familiar one—to describe what and how much is where.

Daniel W. Gade, 1983, p. 265. When researchers avoid topics or regions of great interest to them because they do not wish to make the effort to learn that language, the intention of geography to know the world is unfilled.

Michael C. Steiner, 1983, p. 430. Regionalism seemed to become an American preoccupation during the great depression of the 1930s. During that decade regionalism was widely and urgently discussed by artists, folklorists, social scientists, planners, architects, and engineers. Scores of conferences, roundtable discussions, federal commissions, and symposia were devoted to regionalism, while a swarm of journals sponsored continued debates on the topic. Many of the greatest expressions of American regional thought were published during the 1930s.

Edward J. Taaffe, 1985, p. 97. I differ from several of the regional revivalists in one respect, however: the apparent failure to see a relation between the development of spatial ideas and regional study. I get the impression talking to some regionalists that there is nothing but harm to regional geography in quantitative and scientific work. Actually, there is a great potential enrichment of regional geography in the application of spatial models dealing with diffusion, commuting, systems of cities, migration, and transportation networks to places familiar to regional geographers.

# PART FOUR
# The Spatial Tradition

## SITE AND SITUATION

Robert M. Brown, 1910, p. 121. It is hardly safe to assume that cities and towns owe their existence to physiographic factors alone, for an analysis of these factors cannot explain, except by some process of mental juggling, the location and size of all our cities and towns. I am not going to cast aside the physiographic factor but I am going to advance the statement that the fitness of any site is determined in the long run by the condition, as well as the accessibility of the market.

Mary Dopp, 1913, p. 401. The most important factor in the early history of Wisconsin was its key position between the Upper Great Lakes and the Mississippi River.

Mark Jefferson, 1915, pp. 26–27. Is it not singular that all these cities grow so similarly? Seattle and Los Angeles are 1,000 miles apart and their environments have very strong contrasts, yet each, and all their neighbors, show strong growth in 1890–1900, an immense spurt in 1900–1910.

Mark Jefferson, 1917, p. 6. No city becomes a great one nowadays unless it has a strategic position. All cities are groups of men, in last analysis. Travel and communication have become very easy, and men in every town and city are eagerly competing for the generous rewards of modern business. Any real advantage of position of one town over another must finally bring business that way. Men do not know clearly what position is going to prove strategic, do not always recognize what the advantage of a growing town really is, but the greater growth of one city than of others around it is evidence of some geographic advantage.

Richard Hartshorne, 1927, p. 98. This paper has attempted to outline a method by which we may study the geography of any manufacturing industry to determine the importance of the different factors affecting the distribution of its plants. In general it is found that locus,—i.e., relative location with reference to specific areas—is the all important element; that locus with reference to areas of raw materials, power or fuel, and markets, together with the labor factor are the four major factors influencing factory location.

Mark Jefferson, 1931, pp. 126–127. Life in [the Great City (100,000 inhabitants)] is really urban, which life in towns of from one to five thousand is certainly not.

O. H. K. Spate, 1932, p. 622. A capital may be defined as the place wherein the political authority of a territorial unit is concentrated; it is the seat of the legislature, the headquarters of the executive, exercising a higher or lower degree of supervision over local administration according as the structure of government is highly centralized or federal. It is often, although not invariably, the cultural focus of the country; but in many cases, especially in federal states, its commercial and industrial importance is less than that of other cities in the same country, since the factors that influence the choice of site for a capital are often political and strategic rather than economic. This being so, we may expect to find that any geographical forces which have affected the political history of a country reach their most intensive expression in the choice of site and the development of the capital and are ever reflected in its external appearance.

Robert E. Dickinson, 1934, p. 285. The manufacturing belt . . . is the area of maturest metropolitan development, in which the metropolitan centers are the commercial and cultural focuses for tributary industrial towns.

Meredith F. Burrill, 1940, p. 50. The printing industry is one of those which by their repeated occurrence characterize certain urban zones.

George Kiss, 1942, p. 634. Reduced to its essential elements, Ratzel's theory is built on two main ideas; space (the area occupied by a state) and position. These two determine the geographical value and ultimate destiny of every part of the earth's surface. With the inexorable logic of scientific determinism, Ratzel cites example after example to prove how space and position conditioned the historical fate of an area.

Glenn T. Trewartha, 1943, p. 32. The unincorporated hamlet represents the first hint of thickening in the settlement plasm. It is neither purely rural nor purely urban, but neuter in gender, a sexless creation midway between the more determinate town and country.

Edward L. Ullman, 1954, p. 283. Spatial interaction focuses on circulation and the connections between areas rather than on the nature of areas themselves, although of fundamental importance in explaining the latter. It is closely allied to the concept of situation in geography.

Fred E. Dohrs, 1958, p. 259. Nearly all political geographers, beginning with Ratzel and Semple, have emphasized the importance of location.

Fred Lukerman, 1960, p. 333. The analytic concept ''location'' is one of the least defined terms in the vocabulary of geographical studies. Methodological writers have taken the word as given, or, recognizing its equivocal character, have begged to discuss it at a later date.

Fred Lukerman, ibid. The once sharpest tool of the geographic method—relative location—has become a methodological orphan.

Preston E. James, 1967, p. 6. The adjective "geographic" should refer to anything that is located on the face of the earth, and that is a factor in a problem because of its location.

Preston E. James, ibid. The widespread use of "up and down" to mean north and south is not only irritating, but also it is symptomatic of a deep-seated form of geographic illiteracy. We cannot expect to change the habits of speaking of the general public, including radio and television announcers; yet when we are told that a storm is moving up the St. Lawrence, we may legitimately be puzzled about who should prepare for bad weather. When a geographer gets mixed up regarding the location of the upper Nile, or writes about going down the Rhine to Switzerland, things are in a bad way. The term Sub-Saharan Africa can refer to nothing in the world but the supplies of water and minerals that lie beneath the surface of the Sahara.

Bonnie Barton, 1978, p. 34. The concept of centrality is thoroughly geographic, meant to capture the magnetism of location and the focal character of human activities.

## THEORY AND LAWS IN GEOGRAPHY

Frank Boas, 1887, p. 138. As soon as we agree that the purpose of every science is accomplished when the laws which govern its phenomena are discovered, we must admit that the subject of geography is distributed among a great number of sciences; if, however, we would maintain its independence, we must prove that there exists another object for science beside the deduction of laws from phenomena. And it is our opinion that there is another object,—the thorough understanding of phenomena. Thus we find that the contest between geographers and their adversaries is identical with the old controversy between historical and physical methods. One party claims that the ideal aim of science ought to be the discovery of general laws; the other maintains that it is the investigation of phenomena themselves.

Ellsworth Huntington, 1913, p. 641. ("New Science.") Geography, although perhaps the oldest of the sciences, has changed so completely in recent years that its true nature is not commonly understood, especially in America. Like every other science, it includes three phases or stages, each of which is characterized by a special mental process. The first, or empirical stage, is concerned merely with the gathering of a great body of unrelated facts. In the second, or systematic stage, these facts are classified and arranged in definite categories;

while the third, or explanatory stage, is devoted to the explanation of the facts and to the determination of the laws which govern them. To these may be added the predictive stage, in which the laws of the explanatory stage are used to predict future occurrence; but with this we are not now concerned.

Ellsworth Huntington, ibid., p. 652. The field of geography as the science of the distribution of phenomena upon the earth's surface is distinct and well defined; its laws, although intricate and as yet only beginning to be known, are precise and clear; and its sustained and intricate modes of reasoning are in the highest degree disciplinary.

R. H. Whitbeck, 1926, p. 120. (''Geonomics.'') In industrial geography we are constantly seeking to explain why, for example, certain industries are developed in one region and not in another; why villages grow into cities at certain points and remain mere hamlets at other points; why this place or that some other place is not. In explaining why these things are so, we recognize the outworking of some law. Is it a law of geography?

Langdon White, 1928, p. 365. Although prophesying is undeniably hazardous, the writer feels that the unique combination of favorable geographic and economic conditions for iron-making at Birmingham warrants an optimistic forecast of its future. Certainly no other American iron and steel district is so scientifically located. Straddling its iron ore, coal, and dolomite; commanding the cheapest supply of labor in the United States; having access to the Gulf via a canalized waterway, which offers rates at 80 per cent of its pig iron and 50 per cent of its steel within the district itself, and making the cheapest pig iron in the country, it would appear that Birmingham might expand limitlessly so long as she can market her products at a profit. Accordingly, it is little wonder that many students of the industry look to Birmingham, rather than to Pittsburgh or Chicago (both of which must always have an enormous home demand to supply), as the ultimate development of the United States Steel Corporation.

W. M. Davis, 1932, pp. 214–215. No geographer, therefore, need feel himself unfortunate because of the great diversity of the facts that his composite subject requires him to study; for in the progress of this work he may discover relations and principles which bind his facts together in a thoroughly reasonable manner, and he may then concern himself, especially in his teaching, largely with those relations and principles, and introduce items of fact chiefly to illustrate the principles. Unhappily, geographers are often so impressed with the innumerable facts of their subject that much of their attention is given to individual occurrences in specified localities rather than to the principles which the occurrences exemplify; and this is regrettable.

Mark Jefferson, 1939, pp. 78–79. In most important countries the largest city is two or more times as large as the next largest city, and other cities differ in size by much smaller numbers.

Mark Jefferson, 1939, p. 231. The facts of record the world over seem to justify the law: *A country's leading city is always disproportionately large and exceptionally expressive of national capacity and feeling.*

Ellsworth Huntington, 1942, p. 3. [A] main line along which great progress may soon be expected is plain, old-fashioned locational geography. What sense is there in talking about Thailand to people who do not know whether it is east, west, or south of Burma, or whether its latitude is 5° or 30°. No fact is fully developed in its geographic aspects unless we understand its location in respect to various other types of facts, both physiographic and ontographic. Only thru such comparative studies of location can we reach the main objective of geography in the form of regional descriptions illustrating the laws which govern the relative distribution of the facts of physiography and ontography.

Harold H. McCarty, 1942, p. 282. Despite the central position of the problem of human location in modern geography, no attempt seems to have been made to set down in systematic form the general principles underlying this phase of human activity.

John Wesley Coulter, 1949, p. 61. The principle of the least action, a law of thermodynamics, can possibly, [*sic*] be applied in the field of human geography.

Edward L. Ullman, 1953, p. 57. I recall an economist once telling me that the map was a theory which geographers had accepted.

Edward L. Ullman, ibid., p. 60. If all places were the same, a ridiculous assumption, there would be no area studies and no geography; parenthetically, if the natural environment were the same everywhere, an equally ridiculous assumption, there would still be geography, because of the advantages of specialization and economies of scale, to say nothing of the differences in culture which would create spatial patterns and contrasts. In fact it is instructive to imagine a completely homogeneous natural and cultural setting in order to formulate theories of settlement and occupational distribution.

H. H. McCarty, 1953, pp. 183–184. The desire for a body of theory that could be used to explain characteristics of the economic occupance of various parts of the earth's surface is as old as the interest in economic geography. In detail, it appears that such a body of theory would be made up of sets of interrelated principles, each of which would be designed to explain locational patterns assumed by individual types of economic activities. Locational principles would take the form, "where x, there y."

Fred K. Schaefer, 1953, p. 243. Geography is essentially morphological. Purely geographical laws contain no reference to time and change.

Fred K. Schaefer, ibid., p. 248. As far as they are morphological, these laws are genuinely geographic.

Stephen B. Jones, 1954, p. 111. No young geographers should feel that theory is reserved for balding professors.

Eugene Van Cleef, 1955, p. 108. Geographers must observe every square inch of the earth's landscapes hoping to find enough repetitive cases to enable him [*sic*] to convert hypothesis into law.

Emrys Jones, 1956, p. 373. The laws of social sciences may also be thought of as different in kind from those of the physical sciences. This centers around the belief in free will. According to the belief in an incalculably free will laws can never be found of stringency comparable to those in physical sciences, because the nature of the human mind is such that human actions are not predetermined. The course of human activity cannot, therefore, be accurately predicted.

John Fraser Hart, 1957, p. 164. Geographers need give greater consideration to the value systems of regions. Comprehension of the regional value system is essential to an understanding of the region. More intensive research should enable geographers to establish generic principles concerning the areal variations and relations of value systems. The development of such generic principles would increase the stature of geography as a social science.

William Warntz, 1957, p. 2. It can fairly be said that social scientists have only just begun the necessary search for the basic dimensions of society so essential if principles are to be developed which will permit the unification of the diverse branches of social science through vocabulary equivalences.

Brian J. L. Berry and William L. Garrison, 1958, p. 83. Pick any large area. It will likely contain many small cities, a lesser number of medium-size cities, and but few large cities. This pattern of city sizes has been observed to be quite regular from one area to another.

F. Lukerman, 1958, p. 6. Laws and predictions in science are made only with reference to these models. Scientific knowledge, beyond description, is derived from the testing of models and pertains only to the testing of models. Models serve only as the "universe of discourse" of science, they certainly are not postulates about the natural world beyond our observations. Since hypotheses assumed from observations are testable, in a scientific sense only through the arithmetical (or logico-symbolic) representation of theory, cases for isomorphism—laws of the same form—may be made between sciences with similar models and between theories and mathematical formulae, but hardly between observations and theories.

F. Lukerman, ibid., p. 9. Concepts of equilibrium, equipossibility, minimization of cost/effort, optimum solution, and theories of games, sets and fields are mechanical assumptions presupposing an idealized continuous process or a sequence of determinative results. The use of such concepts in social science assumes complete knowledge of all causal factors present, pre-determined ends (minimization/maximization), and single (point) solutions. The acceptance by geographers of such models to interpret human behavior is certainly less than an empirical approach.

Duane F. Marble and John D. Nystuen, 1958, p. 279. The hierarchical arrangement of shopping centers within a city resembles the hierarchical arrangement of the cities themselves.

Edward L. Ullman, 1958, p. 294. The number of centers in each hierarchical class does not follow the theoretical framework of either Christaller or Lösch. . . . Spacing along routes in many settled areas, however, is fairly regular.

F. Lukerman, 1960, p. 1. ("On Explanation.") I would question Mr. Berry's interpretation of the geographic meaning of explanation, model, and description.

Preston E. James, 1963, p. 98. Theory is a generalization regarding processes. But an important part of the structure of geography has to do with the differences in the degree of generalization. When the face of the earth is examined in detail by direct personal observation, generalizations can be made and verified by the use of directly observable facts.

Preston E. James, ibid. The role of theory differs depending on its purpose. Theory as a guide to the discovery of new theory through research study plays one kind of role: theory as conceptual structure for the purpose of providing geographic insights, [*sic*] plays quite a different role.

Leslie Curry, 1964, p. 146. If the individual decision maker acting rationally be taken as the unit to be considered, conclusions such as "Corn Belt farmers grow corn because they live in the Corn Belt" are obtained but are not very helpful. Alternatively, a history of unique events is written to describe development and the lawfulness or generality of process is missed. Whether history is lawful is a matter for those who write it to decide. That spatial processes occur, to which the historian could not contribute understanding but which are the very stuff of geography, appears self-evident—or at least a worthy article of faith.

Peter Haggett, 1966, p. 277. . . . the sound of progress is perhaps the sound of plummeting hypotheses.

Preston E. James, 1967, p. 1. The construct of a model, per conceptualization, risks dangers of obscuring effective explanation owing to the fact that models normally can originally be only caricatures of reality. The Davison cycle of erosion so obscured progress for a long period by its overacceptance. Social Darwinism, similarly, obstructed effective explanation for a period. The persistence of error becomes an obstruction to progress in geography. The overlong employment of Maury's simple model of the earth's wind systems was such a persistence. Currently the dichotomy of the dualism phrased as physical geography and human geography constitutes an error of persistence.

Preston E. James, ibid., p. 3. . . . no two pinpoints on the face of the earth are identical. Every smallest point of earth-space differs from every other one. Yet in all this diversity there are regularities that can be identified by geographic methods.

Reginald G. Golledge and Douglas Amedeo, 1968, p. 774. Finally, we should point out that acceptance of the ultimate existence of laws is predicated on accepting the principle of rationality or non-randomness in a universe.

Clyde F. Kohn, 1970, p. 212. The emergence of geography as a more abstract, theoretical science appears to have been the most over-riding development in geographical research during the 1960s.

Norton Ginsburg, 1973, p. 2. Moreover, emphasis tends to be on pattern and structure, rather than on those kinetic and processual dynamics which are essential to understanding the functional attributes of complex systems. So-called, "central-place" theory provides a good example of this failing.

D. W. Moodie and John C. Lehr, 1976, p. 134. A theoretical geography is possible, but a theoretical historical geography is a contradiction in terms.

John Fraser Hart, 1982, p. 4. A good theory is nothing more nor less than an idea about the way things may work. It is merely an attempt to impose order and simplicity on complex patterns and processes, and it has little value unless it advances our understanding of the real world.

John Fraser Hart, ibid., p. 12. The would-be revolutionaries identified traditional geography as essentially idiographic, with a heavy emphasis on places and regions. They proposed to replace this approach with nomothetic theorizing about order and pattern. The revolutionary zealots did battle with ideological fervor, and they obviously could be satisfied only by the total destruction of those they perceived as their opponents.

John Fraser Hart, ibid., p. 18. Some of the blame for the breakdown of the traditional model may be attributed to the inability of regional geographers to interpret the results of systematic studies, but most of it seems to stem from the emphasis of systematic geographers on "place" rather than on places. They seem to have been concerned primarily with the abstract notion of place, and have largely ignored the distinctive character of particular places in the real world.

## SCIENTIFIC GEOGRAPHY

Ellsworth Huntington, 1913, pp. 643–644. ("New Science.") Because of the vast number of highly interesting geographical phenomena scattered all over the world, every traveler has felt impelled to gather his own little sheaf. Having published his observations, he has considered himself a geographer, although with no more claim to the title than has the gatherer of a bunch of wild flowers to be called a botanist. Genuine geographers have rebelled against this invasion of their province by men of no adequate training. Yet instead of directing their own energies to the patient sifting of facts, in order to discover laws, they have zealously devoted themselves to mapping new portions of the earth's surface. Their work has been done scientifically and is of great value, but it belongs to the first, or empirical, stage of the science. In their zeal for this work, they have often forgotten the other phases of the subject. Thus, although thousands of men, both travelers and map-makers, have been called geographers, only a handful have given their lives to the work of systematic classification, and still fewer to the final explanatory stage of the science. This, more than anything else, explains the common but fallacious idea that geography is purely descriptive and lacks the qualities of real science.

Ellsworth Huntington, ibid., p. 644. The geographer deals with everything whose distribution can be shown upon a map, whether it be incised meanders, chinook winds, yellow skins, or cowardice. This does not mean that geography is a blanket science composed of interesting bits from all the neighboring sciences, and therefore no science at all, as is sometimes thought. The chemist may legitimately study meanders in the sense of analyzing water, soil, and rock; he may investigate the chemical differences between warm, dry chinook winds and other winds; he may ascertain the nature of the pigments which make one skin yellow and another brown; and he may tell us that the mental state known as fear or cowardice is accompanied by the formation of certain toxic substances which can be detected in the blood or breath, and which hinder the bodily functions. Yet no one thinks of denying that chemistry is a distinct and full-fledged science. The chemist is obliged to study everything, but he does so for the definite purpose of determining the nature and amount of the various chemical elements, their mode of union, and their changes. In the same way the geographer studies everything, but he does so in order to determine where things are located, and why they happen to be there rather than in some other part of the world.

Albert Perry Brigham, 1915, p. 25. Such then is the mode of advance of our science—the old story of interest, hypothesis, test, correction, publication, criticism, revision; progress by error, by half-truth; by zigzag, spiral, and apparent retrograde; by zero-flight, by patient tunneling; some at the salients of progress, and some in the ranks of humble endeavor, the goal in front of all.

W. M. Davis, 1915, p. 67. Two geographers of equal capacity, one an empiricist, the other a rationalist, might have equally careful training, each in his own school, and equally extended experience in traveling over the world; but their resulting mental equipments would not be equal in respect to completeness. The equipment of the empiricist would be limited to concepts formed as the counterparts of observed and generalized facts; that of the rationalist would be extended far beyond the explanatory counterparts of observed facts, by the addition of a great number of possible counterparts of unseen facts, deduced from the general theoretical explanations under which the observed facts belong.

Louis C. Karpinski, 1923, pp. 212–213. The ancient Babylonians contributed a fine start to scientific geography, particularly the 360 degrees in the circle and ideas connected with the location of the position of the stars. The Greeks gave a scientific system of astronomy and of geography, culminating in the geography and maps of Ptolemy. The Arabs made notable advance on the mathematical side, in explorations, in the recording of latitude and longitude, in perfecting and inventing instruments of observation, in the development of tables, and in cartography. The European translators in the twelfth century made the work of the Greeks and Arabs available. Scientific cartography became a matter of national concern first to the Spanish and the Portuguese and later to the Dutch, the English and the French. With the collaboration of astronomers, surveyors, mathematicians, geographers, physicists, and instrument makers, the necessary data for scientific cartography were finally made available at the beginning of the eighteenth century.

Sten De Greer, 1923, p. 2 and pp. 6–7. ("On the Definition.") Geography is the science of the present-day distribution [of] phenomena on the surface of the earth. Essentially, therefore, Geography is not a special but a general science, like Statistics, Mathematics, Philosophy and even History in its widest significance. They concern themselves with the most heterogeneous matters; but in reality they are restricted to the study of a certain abstract quality of the objects or to purely abstract problems.

Each of these sciences has its own method of work, which is quite different from that of the object sciences. The empirical treatment of quantities of Statistics, the rational treatment of quantities by Mathematics, the treatment of logical thinking by Philosophy, the treatment of position or distribution in space by Geography, and the treatment of development and distribution in time by History—all these would seem to be applicable to all kinds of objects, and would seem to some extent to be operations that most men of science need to carry out.

Sten De Greer, ibid., p. 7. It is here relevant to call to mind the long and important period in which Geography has become regarded as consisting partly of a scientific and partly of a humanistic section. Certainly this twofold division will wither and fade away in the same measure as knowledge will be gained of the character of Geography as a general abstract science. Instead of it, there will appear another twofold division, namely into the analytic and the synthetic study of distribution.

Marius R. Campbell, 1928, p. 25. The real work of a scientific geographer is to gather facts, to classify the facts so gathered, to study the bearing of one fact upon another, and finally to establish their relationship to the surface of the earth which must, by definition and by logical reasoning be the foundation of the science of geography.

W. H. Haas, 1931, p. 323. As most of us are aware, the world is not divided into two sets of phenomena, scientific and non-scientific.

S. W. Boggs, 1932, p. 48. A scientific geographical study of boundaries should take into consideration all types of political boundaries—international boundaries, and both major and minor boundaries between administrative divisions within a country.

Salvador Massip, 1932, p. 67. The airplane has completely changed the methods of exploration. . . . The author has made 19 flights over different parts of Cuba, with scientific purposes.

W. M. Davis, 1934, p. 301. ("Submarine.") If any of these inferences seem outrageously improbable at first reading, let the grounds for their outrageousness be deliberately inquired into; and, when they are found, let the necessity of believing them be seriously questioned. What at first appears outrageous may, when those grounds are deep-plowed, possibly come to be regarded as "intrageous"; and the more deeply they are plowed the more reasonably "intrageous" the outrageous inferences may come to appear.

Isaiah Bowman, 1938, p. 1. When we speak proudly of modern technological advances, we are apt to make of them an arithmetical and impersonal sum and call it progress, forgetting that what science supplies is not at all an addition, positive and beneficial, until men have proved it so. The whole of that proof is found in man himself and not in an objective regard for new facts or inventions. Man's degree of safety and happiness is really the algebraic sum of the forces, social and natural, good and bad, with which he must deal, all added to himself.

Carl O. Sauer, 1941, p. 4. ("Foreword.") Particularly depressing has been the tendency to question, not the competence, originality, or significance of research that has been offered to us, but the admissibility of work because it may or may not satisfy a narrow definition of geography. When a subject is ruled, not by inquisitiveness, but by definitions of its boundaries, it is likely to face extinction. This way lies the death of learning. Such has been the lingering sickness of American academic geography, that pedantry, which is logic combined with lack of curiosity, has tried to read out of the party workers who have not conformed to prevalent definitions. A healthy science is engaged in discovery, verification, comparison, and generalization. Its subject matter will be determined by its competence in discovery and organization. Only if we reach that day when we shall gather to sit far into the night, comparing our findings and discussing all their meanings shall we have recovered from the pernicious anemia of the "but-is-this-geography" state.

J. Q. Stewart, 1947, p. 471. The evident tendency of people to congregate in larger and larger cities represents an attraction of people for people that turns out to have a mathematical as well as a merely verbal resemblance to Newton's law of gravitation.

John Wesley Coulter, 1949, p. 61. Human geography . . . has for its subject human beings. It is a social science with its roots, however, in the natural sciences. The student of that discipline tries to answer the question, How have human beings come to be as they are in the world today? Philosophy asks the question, "Why?" Science asks the question "How?" The method of science involves only a few fundamental factors: accuracy in observation, verification of the facts, methods of induction and deduction, and intuition. The method of science was the method of Darwin.

George T. Renner, 1950, p. 17. The Principle of Optimum Location [:] Any industry tends to locate upon a site which provides optimum access to its ingredient elements. If all these elements occur close together, the location of the industry is pre-determined. If, however, they occur widely separated, the industry will be so placed as to be most accessible to that element which would be most expensive or most difficult to transport, and which becomes, therefore, the locative factor for the industry in question.

Edward L. Ullman, 1953, p. 60. Up to now I have not referred to geography as the study of the significance of areal differentiation, the current catholic definition of geography after Hartshorne, Hettner, James, and others. This concept is implied in the spatial or "science of distribution" point of view. I cannot accept areal differentiation as a short definition for outsiders because it implies that we are not seeking principles or generalizations or similarities, the goal of all science.

Fred K. Schaefer, 1953, p. 239. Hartshorne, like all vigorous thinkers, is quite consistent. He does, in fact, reject all social science and is particularly sceptical of the future of sociology.

J. Russell Whitaker, 1954, p. 241. Thomas Jefferson was a well-rounded scientist, and something of a geographer.

Barry N. Floyd, 1962, p. 1. There seems precious little room left for achieving a genuine equality of status between aspiring literary artists and would-be scientists in our field. Hopes for an honorable reconciliation between those who conceive of geography as an art and those who conceive of it as a science now appear dim. To relegate non-mathematical geographers to the teaching of introductory college courses and the training of high school teachers is to fling down the gauntlet with too strong a gesture.

Barry N. Floyd, ibid., p. 2. We are resolved, then, that the artistic and spiritual appeal of geography as the philosophy of place shall not die; the "flowers of art'' must not be weeded out in producing the ''bread grains of science.'' The mathematical artisans should not be permitted to multiply at the expense of the geographical artists, for there is no justification in forcing the humanities into the frame of the sciences just to make geographical research appear respectable to scholars in other disciplines.

Edward A. Ackerman, 1962, p. 294. At the table of scientific leadership in national affairs, where does geography sit?

Edward A. Ackerman, 1963, p. 433. The mind of the scientist, no less than that of the poet or musician, must be structured by thought and experience before it reaches the creative stage.

Richard S. Thoman, 1965, p. 8. Can we subjugate all of economic, urban, and transportation geography—and possibly still other subfields to which certain theoretical approaches have not yet been applied—to location theory? Are we not substituting means for ends? Is location so all-inclusive and all-important in geography that we must return basically to the concept of geography as a science of distributions, a concept which Hartshorne has shown to be inadequate.

Preston E. James, 1967, p. 1. On the present errors of inadequate hypotheses may be built the progress of geography in the future.

Preston E. James, ibid., p. 4. When several artists are set to painting the same scene or even the same person, it is amazing to note the very different ways the same real object is perceived. Similarly the direct observation of things and events on the face of the earth is so clearly a function of the mental images in the mind of the observer that the whole idea of reality must be reconsidered.

Preston E. James, ibid., p. 10. Thomas Jefferson, famed as a statesman, was also among the earliest American geographers. He was concerned to find an explanation of the existence of fossil shells in the mountains high above the sea. But [this] hypothesis was severely criticized by those who insisted on the scientific accuracy of the Bible.

Preston E. James, ibid., p. 24. Whether geography is pursued inductively or deductively, or even whether it is pursued as an art or as a science, is perhaps less important than that it be pursued by honest scholarship, and that the results of this pursuit be communicated.

R. J. Johnston, 1968, p. 575. Geographers have only recently attempted to put their subject onto a proper scientific basis.

Clyde F. Kohn, 1970, p. 214. In the development of geography as a science, the availability of accurate information, aggregated at different scales is crucial. No science can rise above the quality and quantity of data it has at its disposal.

James L. Newman, 1973, p. 22. Starting a hypothesis is becoming a common procedure in geographic writing, but the meaning and function of the term ''hypothesis'' have a wide array of interpretations. Some geographers use the term to pose fairly specific, directional relationships between phenomena, others use it to state more general relationships, and still others equate it with question, explanation, conclusion, contention, assumption, and model. This confused state of affairs is interpreted as the result of scientific faddism and pretentiousness.

Wilbur Zelinsky, 1975, p. 123. In a quite literal sense, science has become the dominant religion of the late twentieth century, effectively displacing the traditional faiths of the supernatural.

Wilbur Zelinsky, ibid., p. 125. The fact that the full corpus of scientific doctrine is open and comprehensible to just a small band of higher priests does not undermine the contention that Science is indeed the dominant religion today of both the elite and the masses. Equally few citizens of medieval Western Europe were privy to the subtler points of Christian dogma. What matters is an almost intuitive grasp by the faithful of the basic ethos of an all-encompassing theology.

Wilbur Zelinsky, ibid., p. 133. Perhaps the greatest discovery of twentieth century science has to do with its own essential nature: that it is, first and always, a social activity, an organized band of human beings who obey the same fundamental rules of organizational behavior as do any other complex group of people, not a disembodied flock of angels soaring unswervingly upward toward the elysian fields of truth.

Anne Buttimer, 1976, p. 277. Strange indeed sounds the language of poets and philosophers; stranger still the refusal of science to read and hear its message. The humanistic geographer, attuned to the voices of scientist and philosopher, cannot afford to dismiss anything which may shed light on the complexities of man's relationship to the earth.

Anne Buttimer, ibid., p. 283. Scientific procedures fail to provide adequate descriptions of experience because of their implicit separation of body and mind within the human person.

J. Nichols Entrikin, 1976, p. 615. Geographers emphasizing an existential phenomenological perspective in their research have referred to their work as humanistic geography. Some humanist geographers argue that their approach offers an alternative to, or a presuppositionless basis for, scientific geography. A review of the philosophies espoused by humanist geographers and an examination of their interpretations of these philosophies suggests that humanistic geography is best understood as a form of criticism.

Marvin W. Mikesell, 1978, p. 10. Although no one has suggested that the epic of the ''new geography'' should be rewritten under the title Paradigm Lost, a revisionist attitude does seem to be evident toward many of the assumptions made in the 1960s, and the science of spatial organization proclaimed then has neither replaced the traditional pluralism of geography nor held the interest (or faith) of all of its protagonists.

W. T. Grvaldy-Sczny, 1979, p. 1. Think of it like this: Geography is riding in a car along with Science and Art. Geography is, in fact, riding in the back seat. Science has been driving for seventy-five years, fighting with Art all the way. Science scorns Art; Art sneers at Science. Neither pays much attention to Geography, except for help with reading the map. Geography tries to take a nap, but cannot sleep. Geography tries to understand why Art and Science fight so, but gives up and looks out the window, which is really more interesting than the fight anyway. Art protests that Science drives too fast. Science snaps back that Art does not understand how to make progress. Geography sometimes sides with Art, often with Science, but neither cares much, nor do either of them care when Geography announces that they are now passing Cleveland. (Science grunts, eyes straight ahead; Art faces so as not to see Cleveland.) Finally, Geography can take no more and tells Science to pull in at the next rest stop.

Peter Gould, 1981, p. 166. Over the past five years, I have tried to find a more satisfying basis for inquiry in human geography in particular, and for human sciences more general.

John Pickles, 1982, p. 388. The rise of positivistic philosophies in the human sciences during the past century and the dominance of methodologies based upon logical empiricist frameworks since the turn of the century, have turned many of the social and behavioral sciences progressively away from their traditional associations with the humanities and with historical concerns. They have adopted the frameworks and methodologies of the natural sciences.

John Pickles, ibid., p. 390. In rejecting positivism, phenomenology does not seek to replace empirical science, but to rid it of the restrictions positivism has placed on what is meaningful and the methods that are appropriate.

John Fraser Hart, 1982, p. 3. We should be quite content—and I, for one, am quite proud—to be recognized as "mere geographers," and it will do us no good to fret about whether anyone else thinks geography is a Science.

John Fraser Hart, ibid., p. 4. The admirable and essential qualities of good science are equally essential qualities of good scholarship, which is a far larger universe than mere science. Science is subsumed under scholarship.

John Fraser Hart, ibid., p. 6. What I am attacking is scientism, the bland assumption that something is automatically superior because someone claims it is scientific, the belief of those who think they will somehow attain higher status and achieve greater prestige if only they can manage to be identified as scientists.

John Fraser Hart, ibid., p. 12. The Quantitative Revolution basically was yet another attempt to make geography "more scientific" by taking it off on yet another new tack, and in the process the revolutionaries tried to force regional geography to walk the plank.

Avinoam Meir, 1982, p. 7. Students who do not learn about the growth and development of geography as a scientific discipline are severely limited in their ability to understand fully its present structure.

Jerome E. Dobson, 1983, pp. 141–142. To speculate on what greater automation means to the future of geography, I believe automated geography will be a major new extension to the discipline. It will draw upon the full spectrum of approaches employed by geographers and will enable us to adopt more techniques from other disciplines. It will be supportive of the scientific and humanistic traditions in geography and other disciplines. It will enable us to compete more readily with other disciplines, such as economics, to address issues of broad societal and technological scope. It will give geographers an improved ability to practice holism as we have claimed it should be done, and it will give the scientific community a new capability to address large, complex problems.

Richard Morrill, 1983, p. 5. The special contribution of geography is not only that our theories, models and analytical techniques enrich and improve the theory and practice of other disciplines, but also that spatial processes, behavior, and differentiation are fundamental phenomena of central scientific importance and comprise the core of the discipline of geography.

David Harvey, 1984, p. 5. The radical and Marxist thrust in geography in the late 1960s concentrated on a critique of ideology and practice within the positivism that then reigned supreme. It sought to penetrate the positivist shield and uncover the hidden assumptions and class biases that lurked therein. It increasingly viewed positivism as a manifestation of bourgeois managerial consciousness given over at worst to manipulation and control of people as objects and at best capable of expressing a paternalistic benevolence. It attacked the role of geographers in imperalist endeavors, in urban and regional planning procedures designed to facilitate social control and capital accumulation.

Shlomo Hasson, 1984, p. 12. . . . humanistic geography represents a duality of reason and feeling, science and humanism. On the one hand, it is a truth-seeking activity, i.e., science, which despite the complex problems of adequacy and validity, strives at attaining a valid knowledge. On the other hand, humanistic geography carries an ethical message which generally implies human freedom and concern for the conditions of human being and place. Humanistic geography thus has to cope with the problem of reconciling scientific understanding and ethics. Insofar as this reconciliation is a goal, the ethic of humanistic geography must itself serve that goal or at the very least seek to identify the salient aspects of this reconciliation.

## SPATIAL PROCESSES

F. A. Carlson, 1930, p. 258. Activity or Change is recognized as a fundamental principle of geography.

Charles C. Colby, 1931, pp. 118–119. The centrifugal forces are defined as those under which certain urban functions migrate from the central zone of a city towards or actually to its periphery, while the centripetal forces are those which attract other functions to the central zone or hold them there.

John W. Frey, 1933, p. 44. Gasoline, the key commodity of the petroleum industry, is a great distance alleviator, and as such, a powerful factor in many geographical relationships.

Isaiah Bowman, 1938, p. 7. Distance, time, and numbers are the elements that in varied combinations set functional limits to those even more varied human aggregates that we call neighborhoods, cities, regions, countries.

Mark Jefferson, 1939, p. 227. Cities grow by excess of births over deaths and by the attraction of opportunity for employment.

Mark Jefferson, ibid., p. 229. The enterprising people who go to the primate cities are pilgrims who purchase no round-trip tickets. They mean to stay.

Carl O. Sauer, 1941, p. 8. (''Foreword.'') The geographer cannot study houses and towns, fields and factories, as to their where and why without asking himself about their origins. He cannot treat the localization of activities without knowing the functioning of the culture, the process of living together of the group, and he cannot do this except by historical reconstruction.

Carl O. Sauer, ibid., p. 9. I am inclined to say that geographically the two most important events of my life-time have been the settlement of the last of the prairie lands and the coming of the Model T Ford, one an end, the other a beginning of a series of cultural processes.

Mark Jefferson, 1942, p. 480. It is normal for great cities, industrial cities, to grow steadily.

Derwent Whittlesey, 1945, p. 9. No statement of geography is complete unless it takes cognizance of the factor of time. In this sense, all geography is historical geography.

Carl O. Sauer, 1952, p. 387. Why indeed should the study of society wish to be anything less than a study of the origins, development, mingling, and extinctions of human institutions and values throughout all human time and among all humankind?

Edward L. Ullman, 1953, p. 56. The geographer's concern with space has been compared to the historian's concern with time, although space may be more concrete.

Rhoads Murphey, 1954, p. 362. The city has been a center of change in western Europe, while it has been the reverse in traditional China, despite the broad similarity in urban economic functions in both areas. Urban character and urban roles may be useful indicators of the nature and dynamics of the diverse entities of society.

Nicholas Helburn, 1957, p. 2. As distances shrink under the impact of modern transportation and communication and as social scientists begin to act upon the realization of the greater interdependence of all parts of the world, there is more and more interest in worldwide classifications of social phenomena.

W. A. V. Clark and Ronald R. Boyce, 1962, p. 14. The implications of brain power as a factor in industrial development are profound. It is quite clear, for example, that the existing aggregations of industry, particularly those located in large metropolises in the American manufacturing belt, have a great advantage over the rest of the nation. Not only are markets, transportation facilities, capital, and the potential for aggregation economies most abundant in these limited areas, but creativity likewise is concentrated here.

Brian J. L. Berry, 1964, p. 2. Geographers are, like other scientists, identified not so much by the phenomena they study, as by the integrating concepts and processes that they stress.

Preston E. James, 1967, pp. 3–4. Although geography is focused on the examination of spatially organized knowledge, the fact remains that space and time cannot be separated, and the time dimension remains an essential part of spatial analysis. In simple language, inherited from the past, "all geography must be treated historically."

Donald G. Janelle, 1968, p. 5. Most scholars in geography and related social sciences have recognized that the locational structure of man's economic, political and cultural activities is not in a static state. Yet very little attempt has been made to conceptualize the manner by which the spatial arrangement of man's establishments change over time.

Gordon R. Lewthwaite, 1969, p. 1. Between the production and consumption of virtually anything and everything lies the intervening phase of preservation and storage, a phase sufficiently important to call for direct and explicitly geographical study.

Clyde F. Kohn, 1970, p. 217. Social geography deals with the social structure and processes of a given society or societies from a spatial or an ecological perspective, just as economic geography deals with their economic structure and processes, and political geography with their political structures and processes. The three together might very well be conceived as parts of a more general "cultural" geography.

J. B. Jackson, 1972, p. 155. There are landscapes in America, separated by hundreds of miles, which resemble one another in a striking degree, and many American towns, even many American cities are all but indistinguishable as to layout, morphology, and architecture. The lack of variety in much of our man-made environment is recognized by anyone who has traveled widely in this country. Many deplore it, try to escape it, and because they cannot, suppose that America is altogether lacking the kind of landscape beauty characteristic of older parts of the world.

William W. Bunge, 1973, p. 332. Historical geography is almost never considered as a useful subject.

Jonathan D. Mayer, 1982, p. 261. The contextual and environmental relations of health and disease provide the basis for medical geography. The challenge of medical geography is that it provides an opportunity for the application of spatial theory, methodology, and understanding to a set of phenomena that are at once the product of biological, cultural, and environmental processes. At the same time, it provides traditional health scientists with a basis for the analysis of pathological processes in their environmental contexts.

Allan Pred, 1984, p. 282. Place is therefore a process whereby the reproduction of social and cultural forms, the formation of biographies, and the transformation of nature ceaselessly become one another at the same time that time-space specific activities and power relations ceaselessly become one another.

## STATISTICAL METHODS AND MODELING

Charles C. Colby, 1936, p. 37. Geographic thought in the coming decades, I predict, will call for much greater accuracy and much deeper penetration than has been true in the past. The application of statistical methods, I believe, will introduce new types of measurement, will clarify present methods of analysis, and will give us results which are quantitatively exact as well as quantitatively true. Here we are on the frontier of geographical thought, for we have come to realize our insufficient preparation for this type of work. Fortunately, some of the younger men are cultivating this type of exploration and we await with high expectation the statistical wonders which we are hopeful they will perform.

Helen M. Strong, 1937, p. 43. It is earnestly hoped that some of the men and women who now are in the beginning of their productive scholarship may make field and statistical studies, geographical analyses and interpretations, which will lead to a better understanding of the regions of manufacturing intensity in the United States, of their close connection with agriculture and other industries, and of the significant contribution being made to the national life by the manufacturing industry.

E. E. Sviatlovsky and Walter Crosby Eells, 1937, p. 240. The development of a fully adequate theory of regional analysis by statistical means is one of the outstanding problems of geography today. The map alone is not enough, nor are statistical data alone. They must be brought together. This is the aim of centrography.

Henri Baulig, 1950, p. 195. Mathematics is an infallible transforming machine, but it cannot return more than it has received.

Kirk Bryan, 1950, p. 208. If, then, we think of geography as at present practiced by the members of this Association, it is in the qualitative stage. The relations of man and his works to his environment are exceedingly complex. Our geographers are attacking their problems by observational techniques based largely on the methods of Natural History. Only in the field of demography are quantitative and statistical methods as yet in use. Many attempts to use statistical methods have met with resistance.

Arthur N. Strahler, 1954, p. 7. If the geographical investigation is quantitative, that is, if an attempt is being made to establish mathematically the parameters of significant regressions between one or more physical factors and some economic or social element, the empirical data are indispensable.

If, in the study of influences of slope on crop production, the investigations were carried out by a geographer who had never studied the explanatory-descriptive methods of geomorphology, he could not make an intelligent selection of the statistics from which the slope data are derived.

If in the training of the geographer the understanding of earth-science processes has been neglected, his later analytical studies will contain inherent defects.

Walter F. Wood, 1954, p. 12. A review of the geographic literature produced during the present century reveals a trend in the direction of more quantitation of findings.

Harold H. McCarty, 1954, p. 227. One of the main purposes of research in agricultural geography is to discover principles that will account for variations in the popularity of individual crops. When these principles are finally perfected it appears that they will consist of equations having variables.

William L. Garrison, 1956, p. 4. Judging from recent publications, computing percentages would seem to be the technique of modern comparative urban geography.

Carl O. Sauer, 1956, p. 298. These misgivings about mapping programs and their techniques rest on a growing conviction that we must not strain to make geography quantitative. Quantification is the dominant trend in our social sciences, which are imitating the more exact and experimental sciences; it happens to be fostered at the moment by the liking of those who dispense funds for long-term programs and institutional organizations. I think we may leave most enumerations to census takers and others whose business it is to assemble numerical series. To my mind we are concerned with processes that are largely non-recurrent and involve time spans mainly beyond the short runs available to enumeration.

Hildegard Binder Johnson, 1956, p. 336. Space is area which is measured by multiplying width and length.

William L. Garrison and Duane F. Marble, 1957, p. 137. The techniques utilized, for the most part set theory, axiomatic structures, and differential calculus, were selected because they have proved to be the most efficient analytical tools available for theoretical maximization problems of this general nature.

F. Lukerman, 1958, p. 6. Quantification of observations does not lift the scientist above description, nor is the describing of observed phenomena in arithmetical language, [*sic*] explanation.

Edward J. Miles, 1959, p. 200. Quantitative data are meaningless when isolated from either their spatial or chronological contexts.

Brian J. L. Berry, 1959, p. 169. One fundamental set of geographic problems involves grouping, classifying, or regionalizing. The common feature of this set of problems is the summarizing of large numbers of individual observations into smaller numbers of groups, classes, or regions, each of which contains members homogeneous with respect to a single criterion or several criteria.

David E. Snyder, 1960, p. 13. A highly flexible method for mechanically sorting data that has been generally neglected by geographers is the peripheral punch card system.

O. H. K. Spate, 1960, pp. 386–387. Humanist as I claim to be, I do not deplore this [quantification] trend so far as it has gone. Quantification has increased, is increasing, and in my opinion ought not to be diminished but to stay—quantitatively just about where it is now.

O. H. K. Spate, ibid., p. 391. This is, like it or not, the Quantified Age. Better to ride the waves, if one has sufficient finesse, than to strike attitudes of humanistic defiance and end . . . in the dustbin of history.

Barry N. Floyd, 1962, p. 1. No, geography must *not* be allowed to lose its fundamental and time-honored attractions as a humane and philosophical field, with esthetic and emotional appeal which lies beyond the cold, rational logic of the mathematical sciences. These are the qualities which drew many of us to the field in the first case, not to mention scholars from countless generations before ours.

Barry N. Floyd, ibid., p. 3. As these words are written, I can almost visualize the condescending smiles on the faces of the quantifiers in our field. Some may even rejoice at the imminent decline of yet another dogmatic geo-artist. But I am also an unabashed geographile; I shall continue to work for the subject that I know and love, whether or not I am obliged ultimately to forfeit the title of geographer.

Barry N. Floyd, ibid. Our plea is for an *emotional* geography, not a soulless, statistically-derived geography. We want a *memorable* geography which will appeal to the hearts and souls as well as to the minds of men. We seek an entertaining, *pleasurable* geography which leaves us visibly moved at the literary skill and brilliance of verbal analysis and filled with wonder for the activities and scenes depicted.

Andrew H. Clark, 1962, p. 237. If our seniors often seem to regard tippling at the statistical inn as a dangerous form of juvenile geographical delinquency, they in turn will be more likely to applaud the exercise of one of the traditionally most highly developed crafts of the geographer which, alas, our quantitative intemperance may have tended to displace a bit from our training and practice. This is the ability to observe, read, and enjoy the urban and suburban landscapes—in brief, in a phrase now fallen too largely into disuse, the face of the earth.

Barry Floyd, 1963, p. 15. Faced with the intellectual immensity of the geographer's goal, a number of young workers in the field would have us eschew the integrative, global philosophy of the subject, time-honored as it is, and turn our attention to sterner, more "masculine" things: a rigorous wrestling with restricted data and areas utilizing the "new" scientific methods, quantitative ways of analysis, statistical and mathematical tools and models, and so forth. The old ways often lead to a never-ending trail of "trivia, boredom and ennui"; the new presents challenging, brain-cracking problems in combinational topology, scaling, theories of central place, location, nearest neighbors, the journey to work, and linear (or at least quasi-linear) programming.

Barry Floyd, ibid., p. 16. The excesses being perpetrated within the field by statistical geographers tempt one to suggest that the cult of quantification represents a deviation from the main stream of geographic thought, and entitles its adherents to join the ranks of those geographers from former times who attempted to foist their concepts rather too vigorously upon the larger fraternity of scholars.

Eugene Van Cleef, 1964, p. 1. An unbiased observer, it would seem, must believe that a research worker in the field of geography should have a familiarity with a more substantial background than the qualitative aspects of his specialty alone.

Reginald G. Golledge and Douglas M. Amedeo, 1966, p. 14. The distinguishing feature of geographical classification is that it attempts to include a spatial element as one of its differentiating characteristics.

Michael F. Dacey, 1966, p. 555. The probability that a point is within $x$ units of its equilibrium position on the $X$ axis is

$$F(x) = \frac{1}{\sigma_x(2\pi)^{1/2}} \int_{-x}^{x} e^{-x^2/2\sigma x^2 dx},$$

and the probability that the point is displaced $y$ units from its equilibrium position on the $Y$ axis is

$$F(y) = \frac{1}{\sigma_y(2\pi)^{1/2}} \int_{-y}^{y} e^{-y^2/2\sigma y^2 dy}.$$

Wayne K. Davis, 1966, p. 300. In recent years a much needed framework for geographic research has been provided by systems analysis.

F. Hung, 1966, p. 343. The mathematization of economic geography appears to be only at its beginning.

Raymond E. Crist, 1969, p. 305. Cultural geography is fluid, like the subject matter with which it deals. We cannot tether all the accumulating geographic material to the New Geography, like a goat to a post. Geography is geography.

Clyde F. Kohn, 1970, p. 213. The development of a more theoretical geography during the past decade has also led to a concentration on model building. I do not refer here to iconic models or analogues which have been used by geographers for a long time, but rather to statistical and mathematical models in which the spatial or ecological properties of the relationships represented are expressed symbolically. These models become useful as they aid in the accumulation of natural and cultural events, or of the ecological relations of natural and cultural phenomena.

Norton Ginsburg, 1973, p. 2. The new orthodoxy demands the analysis, that is, the breaking down into as small units as possible, of extremely complex systems. Measurement and quantification, those nineteenth century reductionist consequences of social Darwinism, become the fetishistic *sine qua non* of acceptable scholarly enterprise.

William W. Bunge, 1973, p. 332. Quantification and theoretical geography are being attacked as antihuman. This is being done in spite of the content of much mathematical and theoretical geography whose humanist literature is impressive. It is true that some mathematical and theoretical work obfuscates the human condition, but this is a matter of misapplication rather than of the intrinsic inhuman quality of the work.

James Parsons, 1977, p. 1. The geography that has most appealed to me has been a kind of historically based "landscape appreciation" and I confess a certain uneasiness with the current compulsion for precision of analysis, the often sterile straining for statistical content and significance.

William L. Garrison, 1979, p. 120. The question I am asked most often is, "what stimulated quantitative work during the 1950s?" Partly, it was the ambience of playing with ideas, and partly, for me, it was disenchantment with my Ph.D. thesis which imputed causality from taxonomic evidence. But there were many contributing factors. Torsten Hägerstrand spent an academic year with us, the Hartshorne-Schaefer debate was on. Several of the ingredients were in the department for the whole decade. There, the paradigms of regional and systematic geography were in clashing encounters; paradigms for the study of urban areas, transportation, and regional science were accosting one another. Confrontations bubbled up, and we played with methodological ideas. And the accomplishment at the University of Washington during the 1950s was the production of a generation of students who liked to play with ideas about the logic of method.

John Fraser Hart, 1982, p. 17. Half the people who hold doctorates in geography have received them within the last ten years, and three-quarters within the last twenty, in what without distortion might be described as our Quantitative Era, when geography seems to have emphasized techniques and procedures rather than places and people. Many of these younger geographers have not been able to develop a defensible philosophy of geography because they have not been given a sympathetic exposure to some of the grand traditions that have helped to give the discipline a sense of unity.

Shlomo Hasson, 1984, p. 12. In studying people's relationship with place, humanist geographers wish not to confine themselves to quantifiable and observable phenomena. Rather, they try to interpret people's values and purposes, clarify the embodiment in space of such values and purposes, and elucidate the social meaning of space.

## SPATIAL ORGANIZATION

C. R. Dryer, 1911, p. 107. Facts of location, direction, distance, area and distribution belong exclusively to geography. Distribution is not simply static, as shown on a map, but dynamic as a result of terrestrial activity. Hence geography becomes a science which investigates the general laws of distribution of every class of phenomena on the earth.

Emory R. Johnson, 1910, p. 243. To understand the freight services of American railroads one must have at least a general knowledge of the sources of traffic.

Michael Haltenberger, 1915, p. 729. The Egyptian obelisk which adorns Central Park, New York, is one of the few objects that represent in their history phases of both ancient and modern methods of transportation. Hewn from the syenite of Assuan, it was transported over 600 miles to the present Alexandria, where it stood for many centuries. No one knows how many years, how many thousands of men were spent in the slow, tremendous labor of dragging the huge monolith from Upper Egypt to the edge of the Mediterranean.

Then came a time when it was loaded in the hold of a steamship and in a brief period was transferred by water and up through Manhattan Island to its present place. Mechanical science made this final journey, fraught with some difficulty, to be sure, a mere bagatelle as compared with its first removal many ages ago.

Raymond E. Crist, 1932, p. 422. But the rich resources of grasslands and forests, of plain and mountain [*sic*] will not be realized until the transportation problem, the common problem of so much of South America, is solved.

J. Russell Whitaker, 1932, p. 164. A matter of first importance in geographical science is the interdependence of areas. It is possible to describe an area with little regard for the lands surrounding it, but it is much more difficult, if not impossible, to explain how the area came to its present development without considering the relationships with other parts of the world. These relationships appear to be almost wholly dependent on phenomena of a *circulatory* nature—on the movement of winds, waters, animals, plants, etc., as well as of goods, men, and information.

J. Russell Whitaker, ibid., p. 165. Perhaps the principal thread connecting regions in a symbiotic group like the snails, fish, frogs, and green plants in a "balanced" aquarium, is trade.

Theodore Roosevelt, 1934, p. 182. In my own opinion the most satisfactory condition for a community of this sort is to have its land divided into reasonably small farms owned by individual farmers. Obviously this is not the method whereby the greatest efficiency can be attained, nor the greatest gross wealth for the nation. Large plantations owned by single individuals or companies and run by laborers are normally administered in a more efficient manner and can with greater ease avail themselves of market facilities, financing where necessary, and modern agricultural methods. On the other hand, the assets coming from small individual holdings more than offset this. Though the gross wealth may be greater where the land is divided into large holdings, the wealth of the average citizen in a community of small farms is greater. As a result, he has a greater stake in the community and is normally a more thoughtful and more contented individual. Furthermore, when a man owns his own farm there rests on his shoulders a responsibility that develops his ability and character.

J. E. Switzer, 1936, p. 99. The purpose of this paper is to point out some of the indications of the dependence of peoples upon each other, tho they may be culturally and geographically widely separated.

Edward A. Ackerman, 1938, p. 253. No railroad exists in Albania, and the principal stretches of pavement on its recently "rehabilitated" highways are composed of stones laid by the Romans. Pack-mule is and for some time will remain the principal means of transportation.

Robert E. Dickinson, 1938, p. 609. From the dynamic point of view an economic region is an area of interrelated activities and kindred interest and organization. It is an entity of human space-relationships, which are effected through the medium of the route pattern and the urban centers. Such a region, therefore, embraces the complex and closely woven fabric of intercourse by means of which are effected the transfer of goods and persons and the distribution of services, news, and ideas, the very bases of society.

Carl O. Sauer, 1941, p. 6. The ideal formal geographic description is the map. Anything that has unequal distribution over the earth at any given time may be expressed by the map as a pattern of units in spatial occurrence. In this sense geographic description may be applied to an unlimited number of phenomena. Thus there is a geography of every disease, of dialects and idioms, of bank failures, perhaps of genius. That such form of description is used for so many things indicates that it provides a distinctive means of inspection. The spacing of phenomena over the earth expresses the general geographic problem of distribution, which leads us to ask about the meaning of presence or absence, massing or thinning of any thing or group of things variable as to areal extension. In this most inclusive sense, the geographic method is concerned with examining the localization on the Earth of any phenomena.

Glenn T. Trewartha, 1943, p. 55. There does not seem to be nearly the same avoidance by hamlets of main-traveled highways that there is of rail lines. There is some evidence indicating that hamlets tend to avoid areas in close proximity to larger settlements and to become more numerous in the inter-village and inter-town districts.

Erwin Raisz, 1944, p. 81. Our earth is regular in shape only; almost all geographically interesting features have a very uneven distribution. Most land is crowded into one-half of the globe, and population, and all that goes with it, is crowded into certain corners of this land. This unevenness of global distribution is an important geographical factor, especially worthwhile to study at the present time.

Kirk Bryan, 1944, p. 185. Time, thought, and money are never wasted on the production of a map which shows clearly the relation of one thing to another on the earth's surface.

Wallace W. Atwood, 1947, p. 11. All physical barriers that helped to make isolation possible in the past have been broken down. Science and technology have nearly removed the factor of distance in travel and completely removed that factor in communication. It is impossible with modern means of travel and communication for any except the most primitive and illiterate or enslaved peoples to be shut off very long from information about the other parts of the world or to resist the temptations of trade and intercourse with other peoples.

Edward L. Ullman, 1949, p. 19 and p. 21. An entirely different way of achieving results would be to adopt a method somewhat like that shown on the map of *British Empire Shipping, 1936*, which shows the position of all British ships on March 7, 1936. Each dot on the map represents a British ship.

Edward L. Ullman and Walter Isard, 1951, p. 179. The scope of economic geography should be extended to embrace the study of flow phenomena, heretofore neglected in both economics and geography. The effective study of flow phenomena requires a new method of analysis which necessitates some of the concepts and techniques of both sciences.

Preston E. James, 1952, pp. 200–201. The phenomena which produce likeness and differences between places on the face of the earth, whether static or kinetic, form from different kinds of patterns. These are patterns of lines, patterns of points, patterns of areas, and patterns of volumes.

Edward L. Ullman, 1953, p. 56. By spatial interaction I mean actual, meaningful, human relations between areas on the earth's surface, such as the reciprocal relations and flows of all kinds among industries, raw materials, markets, culture, and transportation—not static location as indicated by latitude, longitude, type of climate, etcetera, nor assumed relations based on inadequate data and *a priori* assumptions. I do, however, include consideration, testing, and refining of various spatial theories and concepts.

J. Wreford Watson, 1953, p. 313. Geography has been defined as "the science of space in relation to the earth." Whether geography is all that is debatable, but at least that is all geography.

J. Wreford Watson, ibid., p. 316. Is geography, therefore, simply the spatial factor in other disciplines, and if so, could it not be attended to by them when they come up against it?

No geographer would be happy to answer in the affirmative, just as no historian would be willing to yield up the study of the time factor to any and every subject. These factors of space and time require special study in themselves.

J. Wreford Watson, ibid. Geographers then cannot give the space factor to other disciplines. This is their special study.

Edward L. Ullman, 1954, p. 283. Measurement of actual interaction is difficult because of the sparsity of quantitative data on origin, destination, route, and direction of flows of goods, energy, people, or messages. It is often desirable, therefore, to check interaction on the basis of results, such as location of factories, cities, or trade areas. Many of the assumed interactions or interrelations in geography, however, are modified when actual measures are obtained; yet, no one measure serves for all interactions; people, goods, and messages may not move in like patterns.

Chauncy D. Harris, 1954, pp. 315–316. Individual industries vary, of course, in their locational requirements both in respect to processing costs and to transfer (transport) costs. The location of some factories is strongly affected by regional differences in processing costs, either of labor (cotton textiles) or of power (aluminum reduction). In order to minimize total transport costs other factories are best located between sources of raw materials and markets. Factories that sharply reduce either the bulk or perishability of the materials in processing minimize total costs by locating near the source of raw materials; such are ore concentrating plants, sugar beet factories, creameries, cheese factories, sawmills, and canneries. These factories constitute only a small and decreasing fraction of manufacturing. Factories greatly increasing either the bulk or perishability of products locate near *local* markets: bread bakeries, ice cream factories, ice works, gas works, bottling plants for soft drinks, building construction, newspaper printing. Such industries, though large in total volume, are ubiquitous and do not contribute substantially to regional differentiation.

Rhoads Murphey, 1954, p. 349. Every sedentary society has built cities, for even in a subsistence economy essential functions of exchange and of organization (both functions dealing with minds and ideas as much as with goods or with institutions) are most conveniently performed in a central location on behalf of a wider countryside. The industrial revolution has emphasized the economic advantages of concentration and centrality.

Eugene Van Cleef, 1957, p. 2. The focus of attention upon cities by American geographers is so recent, the year of its beginning can be pinpointed. F. V. Emerson's "Geographic Interpretation of New York City," derived from his Ph.D. thesis written at the University of Chicago and published in 1908–09, may be viewed as the first in the recent period of urban geographic research in the United States.

Paul B. Sears, 1958, p. 16. Our future security may depend less upon priority in exploring space than upon our wisdom in managing the space we live in.

F. Lukerman, 1958, p. 8. Time and distance in social science are not referent to the same kind of area of phenomena (facts) as in physical science.

Allen K. Philbrick, 1958, p. 285. One of the outstanding characteristics of the patterns of world cities is their concentration in clusters and groupings within major urbanized countries.

Charles W. Boas, 1959, p. 170. In a geographical sense, the manufacture of passenger automobiles is most often associated with Michigan and the city of Detroit.

Stephen B. Jones, 1959, p. 252. Thomas Jefferson played a leading role in applying geometry to the American landscapes.

Clyde F. Kohn, 1959, p. 123. Three major considerations are commonly included in the study of human activities in space: area extent, areal association, and spatial interaction.

Edward A. Ackerman, 1963, p. 436. The basic organizing concept of geography has three dimensions. They are: extent, density, and succession.

Edward A. Ackerman, ibid., p. 437. [An] important characteristic of a system is the existence of subsystems within it. The pretty girl, if you like, can be broken down into an astonishing number of subsystems, like any complex being.

William D. Pattison, 1964, p. 211. Entrenched in Western thought is a belief in the importance of spatial analysis, of the act of separating from the happenings of experience such aspects as distance, form, direction and position.

William D. Pattison, ibid., p. 212. A review of American professional geography from the time of its formal organization shows that the spatial tradition of thought had made a deep penetration from the very beginning. For Davis, for Henry Gannett and for most if not all of the 44 other men of the original AAG, the determination and display of spatial aspects of reality through mapping were of undoubted importance, whether contemporary definitions of geography happened to acknowledge this fact or not.

William Bunge, 1964, p. 28. Many basic spatial problems are of a classic dialectical nature. For instance, consider the proposed bridge across Puget Sound to Bainbridge Island in the Seattle-Tacoma Region of Washington. Should such a bridge be built? If the people now living on the island must generate all the traffic, it is obvious that there would not be enough traffic to warrant the construction of the bridge. However, it is immediately apparent that the proposed bridge would greatly increase the population on the island and then there might be enough traffic to warrant construction. Which generates what, the traffic the bridge or the bridge the traffic? Other areas of geography produce similar logical problems. Does the pattern of the valley cause the river to flow? Does the location of "resources" locate the people, or are the opposites true?

Robert Alexander, 1964, p. 1. Space science has been called "the collection of scientific problems to which space vehicles can make some specific contributions not achievable by ground-based experiments."

Brian J. L. Berry, 1964, p. 3. The integrating concepts and processes of the geographer relate to spatial arrangements and distributions, to spatial integration, to spatial interactions and organization, and to spatial processes.

Charles Edwin Williams, 1965, p. 656. There is little information available concerning spatial distribution of fur production.

Preston E. James, 1967, pp. 3–4. Although geography is focused on the examination of spatially organized knowledge, the fact remains that space and time cannot be separated, and the time dimension remains an essential part of spatial analysis.

Preston E. James, ibid., p. 4. It is the geographer's quest to search for significance among all the complexities of the face of the earth. What is significance? To the geographer significance means the identification of regularities and functional interconnections within spatial systems. Because of the trained interplay between concept and percept, the geographer can discover meaning among the things and events that differentiate the face of the earth, much as the artist can lead people to see beauty where beauty was not perceived before.

Stephen C. Jett, 1970, p. 662. As far as I am aware, no comprehensive, worldwide consideration of the blowgun's forms, evolution, dispensals, and distributions has heretofore been attempted.

D. W. Meinig, 1972, p. 159. The four cardinal points—north, south, east, and west—are heavily charged with meaning in the American regional consciousness. Such terms do not denote simple grand quarters of the contiguous national territory but rather are used most commonly as pairs—North and South, East and West—referring to vaguely defined "halves" existing in some degree of contrast and tension.

Robert David Sack, 1972, p. 61. Questions are the mark of a discipline, and the core of geography's questions concerns properties of location: "Where are things?" "Where will they be?" "Why is it here rather than there?" Even questions about a single place are often couched in terms about other places: "What is the connection between this place and that?" "Why is it here?" To the extent that geographic questions are about events in a space, they are about geometric properties and are aptly termed "spatial" questions.

Edward J. Taaffe, 1974, p. 8. On the positive side, the spatial view seems to have been more productive of cumulative generalizations than its predecessors. As the studies progressed, the emerging generalizations showed a certain amount of consistency and cumulation. The findings of one study were starting to be used in the next study. In the late fifties much attention was given to classical theories and ideas related to spatial organization. These earlier statements were subjected to considerably more rigorous empirical tests than had previously been the case, and were reformulated for further testing.

Edward J. Taaffe, ibid. One of the most significant positive aspects of the spatial view has been disciplinary overlap. Other disciplines have subfields involving the spatial expression of phenomena which clearly overlap certain topical subfields of geography.

Edward J. Taaffe, ibid., p. 9. Far from creating a problem, however, these zones of overlap between geography and the social sciences have been key areas of realistic interdisciplinary cooperation during the sixties.

Anne Buttimer, 1976, p. 284. In many respects geography and phenomenology have arrived at similar conclusions about the experience of place. The routes of their investigations are different, however, and hence they offer valuable critical insight to one another. The phenomenologist notes that a social scientist using a priori disciplinary models to investigate experience may fail to tap direct experience. The social scientist may object to the tendency in phenomenology to universalize about human experience from individual accounts. A geographer would be justifiably skeptical about some of the generalizations which have been propounded about lived space. The ideal person described by phenomenologists appears to be rural (at least "local") at heart; nonplace-based social networks do not seriously influence his knowledge of space, or his attractions or repulsions from places. Surely a person could be psychologically present in distant spaces and milieux: places inhabited by loved ones, or milieux rendered vivid through literary or visual media. Does "home" always coincide with residence? Could a person be "at home" in several places, or in no place? Could the gestalt or coherent pattern of one's life space not emerge from mobility as a kind of topological surface punctuated by specific anchoring points?

Julian Wolpert, 1976, p. 13. By what criteria shall we measure and evaluate our communities but by their becoming communities? The alternative to community is confinement.

Stephen Gale, 1977, p. 267. During the 1950s and 1960s geographers made a great deal out of the "geography as geometry" perspective on the discipline. The "spatialization" of disciplinary thought came to be regarded not only as geography's ultimate intellectual and theoretical justification, but somehow its rightful place in the organization of science.

Richard Morrill, 1983, p. 5. What is special or unique about geography is simply that its object of analysis is the earth's surface, and that its purpose is to understand how that surface is structured or differentiated.

# PART FIVE
# Other

## APPLIED GEOGRAPHY

Anonymous, 1813, pp. 234–235. It requires but a slight acquaintance with the conversation of our ephemeral politicians to discover the gross ignorance which prevails among them of geography, and the laughable errors they are perpetually falling into for this reason. Were this ignorance confined to commonplace politicians alone, it might be endured, but that it should discover itself in those who have been honored with the appellation of statesmen is scarcely sufferable. A ludicrous blunder, made a few years ago by an honorable member of congress, is still fresh in the minds of many. A bill was introduced in which some circumstances respecting our trade to India were involved. After some debate in the house, he arose with apparent emotion, and expressed his astonishment that the house should manifest any disposition to enter into the mediated measures, and gave as an insurmountable objection the impossibility of our vessels passing to India without being molested by the Algerine cruisers. Whether the honorable gentleman thought that a canal had lately been cut through the deserts of Suez or that the Algerines lived in the West Indies, South America, or the west coast of Africa is impossible to conjecture; it is sufficient to add that he was interrupted in his speech by a general burst of laughter from every quarter of the house.

Eugene Van Cleef, 1915, p. 138. Geography in the schools probably never has enjoyed the wholesome respect of the business man. Yet this man is in reality not averse to it. The geographic knowledge of his employees not always has satisfied him; logically he has attributed the deficiency to the school training. On the other hand, he probably never has stopped to inquire into the status of geography in the schools. It is not difficult to convert the business man if you can show him "value received." That geography is a school subject that may bring returns cannot be doubted. It seems possible that it may establish itself much more firmly than ever through the medium of the business man.

Wallace W. Atwood, 1919, p. 71. The war has emphasized the need of more trained geographers in America. The Army has called many into service, the Shipping Board has taken others. Some have been sent to France to act as experts;

others have served in this country, either in the training of officers or in the organization of geographical data of special significance to the Army, the Navy, and the Department of State. There appears to be a great awakening in this country to an appreciation of the value of advanced studies in geography. Those who attempt to represent us abroad should know the geography of this country, and should understand the influence of geographic conditions upon the peoples with whom they must deal. We must know better the peoples of the world, and to accomplish that we must know better the geography of the world.

Harlan H. Barrows, 1923, pp. 5–6. History . . . deals largely with the past. Geography proper deals with the present. The historian . . . begins his studies with what our remote ancestors saw; the geographer begins with what we ourselves see.

Helen M. Strong, 1929, p. 49. Only the professional geographer can provide business with this working equipment, for he alone knows how to obtain, assemble, correlate, interpret, and apply scientific geographic material. As large business enterprises become familiar with the value to them of this geographic contribution, they will employ geographic advisers and seek people with geographic training when they add to their staffs. These geographically trained men and women will enable American business to plan its campaigns with an actual world picture in mind, visualizing definitely the lands into which their salesmen and goods are going. They will guide American enterprise with scientific statesmanship at home and in every part of the globe.

Isaiah Bowman, 1932, p. 93. If geography is ever to influence political and social policies it must deal with ideas that seem to be of critical importance to government and society, conveyed in terms that leaders can understand. May I suggest that every person and every subject goes through what might be called a technological phase. Youth enjoys the sound of new terms. Who has not liked to play with dot maps and land-use maps, with the symbols of cartography, with surveying instruments, with the detailed analysis of land forms? All these are extremely valuable tools; but for what? Surely for the purpose of serving ideas that are believed to affect the progress of science or the destiny of mankind. Otherwise technology remains a game and often a kindergarten game at that.

Mark Jefferson, 1933, p. 47. (''Communications.'') Communications are regarded as involving the most effective of man's tools that have fostered his growth in civilization.

Sidman P. Poole, 1935, p. 51. Governmental planning movements offer varied possibilities for Geography and the geographer. Geography may become merely the handmaiden of political cliques. Or on the other hand, the trained geographer may be the man called upon to make the recommendations in these matters of State planning rather than the lawyer or the sociologist.

Isaiah Bowman, 1938, pp. 2–3. Geographers, like all other scientists, have tended to follow fashion, and the fashion now is the capacity of a land for people under certain assumptions. Many geographers may therefore be called capacitators, if I may coin a term. Fewer still are philosophers. The more they learn about men, the less inclined they are to generalize about mankind. Unhappily, they have their share of technicians who merely work a scheme or follow an arid classification based on a priori definitions or speak the patterned jargon of the day.

Eugene Van Cleef, 1944, p. 242. Students who have had courses in geography can apply their knowledge to gardening, especially in these times of food scarcities.

John K. Wright, 1944, p. 199. Two qualities are requisite for successful research: enthusiasm and technical proficiency.

Richard Joel Russell, 1945, p. 301. The public has become extremely geography conscious during the war years. Geographers have enjoyed unprecedented employment opportunities. Their talents have been exhibited in circles hardly aware of their existence less than five years ago. An increased emphasis on geography in school curricula is already evident and everyone expects the movement to gain momentum. The fond hope of geographers, that our subject be viewed in the perspective of its real value, has been realized.

Edward A. Ackerman, 1945, p. 121. Scholars and administrators who had scarcely heard of geography before Pearl Harbor are now familiar with its methods and results.

Preston E. James, 1946, p. 279. Ever since the days of the ancient Greeks, geography has been an essential handmaiden of government. There was need of accurate geographic knowledge before plans and policies could be adopted by the administrators of countries, or by the generals of armies. There has been no question down the centuries about the need for geographical studies at the professional level. But the popular interest in geographical matters has waxed or waned in proportion as the people were called on to share in the making of policy or as they were personally concerned with events in distant places. During both World Wars the American public showed a distinct increase in its demand for geographical knowledge, and especially during World War II the popular presentation of geographic facts thru the use of maps in magazines and newspapers has reached new heights of both quantity and quality. Naturally, therefore, there is an insistent demand for more geography, taught by well-trained teachers in the elementary schools.

William Applebaum, 1947, p. 1. For many years it has been the custom for those who majored in geography to seek an outlet for their professional energies in the teaching world. A few have gone into government service. On rare occasions a venturing soul sought to attach himself to the hardy business world.

Charles H. Mason, 1948, p. 104. Throughout history geography has been the basis of all calculations in war.

John Kerr Rose, 1954, p. 1. Government service now engages something like 500 geographers and is exceeded only by the long established field of teaching as an employment area.

Harold M. Mayer, 1954, p. 7. The relationship between the geographer and the planner have become very close, and a significant number of persons trained and experienced in both regional and systematic branches of geography have found, and are finding, satisfactory and stimulating employment with official agencies concerned with city and regional planning.

Alan S. Gardner, 1954, p. 208. The geographer who is interested in the application of his academic training is able to contribute a variety of unique opinions and concepts to the various levels of marketing research activity. Many of the problems of day to day business research center around geography as such and actually include the fundamentals and the theories of physical and economic geography.

Joseph A. Russell, 1954, p. 279. Every businessman is functioning as a geographer when he makes decisions such as those concerning manufacturing or store location, but few geographers have devoted themselves to assisting the businessman to make *better* geographic decisions.

William Applebaum, 1955, p. 164. Geographers who have done no work for business should indicate their interest in such work.

William Applebaum, 1956, p. 3. Today we are adding to the still-infant literature of our field. This . . . should convince us that there is more pay-dirt in our midst. Perhaps we are ready to start a little mining operation. Certainly we can step up our efforts to improve our professional skills and prepare to bring more forcefully to the attention of business the service that geographers can perform.

Eugene Van Cleef, 1957, p. 2. The geographer who would concentrate his attention upon the geographic without taking into consideration the nongeographic, [ *sic*] courts trouble.

Gilbert F. White, 1962, p. 279. The contributions which geographic thought can make to the advancement of society are relatively few, simple, and powerful. They are so few and simple that a significant proportion of them can be taught to high school and beginning undergraduate students. They are so powerful that failure to recognize them jeopardizes the ability of citizens to deal intelligently with a rapidly changing and increasingly complex world.

Edward A. Ackerman, 1962, p. 292. Geographers are not alone among scientists and other scholars in being conscious of public policy, but it is not a field to which we have come lately. We were interested in public policy long before the great cloud of scientific concern for it mushroomed at the end of the Second World War. I need only cite the application of political and economic geography to the problems of the peace at the end of the First World War, the application of geography to land use and water development planning after 1932, and the application of geography to design of our foreign intelligence operations after 1941.

Edward A. Ackerman, ibid., p. 298. We shall be influential, or capable of being influential, on public policy in proportion as results flow from our fundamental research. We shall be strong and capable if our fundamental research is dynamic, searching, attractive to the best young minds, and in tune with the advancing fronts in other sciences.

Homer Aschmann, 1962, p. 292. Geography's potential contributions to the welfare of the society on a theoretical as well as an applied level are major. Their realization is our challenge.

Eugene Van Cleef, 1964, p. 1. The problem of the geographer is not unlike that of the city planner.

Sherwin H. Cooper, 1966, p. 1. In planning, geography has discovered a new, but philosophically troublesome friend. Planning does not share with the theoretical and historical sciences the collective search for knowledge and understanding of past and present reality. Its sole commitment is to the practical human end of what should be. . . . It is, in other words, a practical science as opposed to a theoretical science, employs normative, rather than empirical premises, and fulfills social, rather than scientific goals.

Nat J. Goode, Jr., 1968, p. 396. The retail geographer is uniquely qualified to provide data on proposed new store locations.

Richard V. Francaviglia, 1971, p. 500. Geographers have published relatively little on necrogeography.

Norton Ginsburg, 1973, p. 4. On occasion, geographers have been articulate spokesmen on major policy issues in this country and abroad, whether one thinks of regional development and water resources policies in the United States, the transformation of the Puerto Rican economy, the reorganization of Europe after the First World War, and the mobilization of academic resources for intelligence activities in the Second; but now, even if the voice of the geographer is heard throughout the land, it is too faint for most policymakers to hear it.

William W. Bunge, 1973, p. 295. ("Human Survival.") To survive, we must learn to worship children, not machines.

Richard Peet, 1975, p. 277. ("Geography of Crime.") Geographers have a legitimate desire to make their discipline "useful." The questions are, however, useful to whom and for what purpose. In the nineteenth and early twentieth centuries geography in the form of exploration, description, mapping and cataloguing was an obvious tool of the imperial powers. Now liberal geographers interested in the study of social problems wittingly or unwittingly serve the interests of today's monopoly-capitalist state by conducting research in the management of problems rather than their solution. In doing so, they protect the system from the full savage force of its own inner contradictions, as essential a service for the survival of the existing order today as imperialistic exploration was in the past.

Julian Wolpert, 1976, p. 5. Our individual skills are the community's resources, not ours, and by analogy individual skills and talents do not exist except in a community context. Loud applause needs the hands of many. The excluded do not applaud. Communities need their clients within their own ranks, just as clients need a community, but empirically concerned acceptance of responsibility and volunteerism are a function of scale both in time and in space. The prideful statement "we take care of our own" has no virtue or meaning without specifying the range of exclusiveness and inclusiveness.

Stephen Gale, 1977, p. 267. With the geometric perspective . . . , geography became removed from social and behavioral problems.

John W. Frazier, 1978, p. 233. Applied geography has a long history, but the new trend differs from tradition. First, it is based in the "new geography," an unprecedented concern for rigorous method and a strong emphasis on human geography. Second, this trend is resulting in major changes in geography curricula in the United States.

G. S. Dunbar, 1978, p. 239. My own opinion is that "applied geography" is a rather unfortunate term. It might imply the existence of a non-applied, inapplicable, geography, or perhaps even a useless geography.

Philip R. Pryde, 1978, p. 1. Geographers talk about planning, write about planning, complain about the failures of planning, extend advice to planners, offer courses to train planners, and speak about the relevance of geography to planning; but in the final analysis, relatively few geographers have any actual experience in planning. If geographers want planners to take them seriously, they must demonstrate they can play the game; they must become, to a certain extent, planners.

Brian J. L. Berry, 1980, p. 449. As professional geographers we share in a common task, that of creating future geographies.

Brian J. L. Berry, ibid., p. 453. We must expand rather than deny our commitment to and reliance upon the world of practice—what others, more felicitously, term the world of the non-teaching professional—for this world must be to the explorers of future geography what field observation was to the geographical exploration in times past.

R. J. Johnston, 1981, p. 213. Geographers, like most other scientists, have actively promoted their work as having an applied as well as an academic role. The reasons why they do this are several; they range from altruism to personal aggrandizement, and from attempts to create major changes in one or more aspects of society and its environment to strenuous efforts to protect the *status quo*. In general, geographers have probably had less impact on public policy than have other scientists, although clearly some individuals have been very influential and others wish to be so. The desire to be "relevant" is strong among academics, especially during a time of economic stringency when research funding is difficult to obtain and sponsors are increasingly determined to invest in work which promises more than an academic pay-off.

Jerome E. Dobson, 1982, p. 137. As automation improves our ability to conduct holistic analysis, the methods of geography will be applicable to a broader range of problems, many economic issues will be seen more clearly as a subset of the total geographical problem, and geographers will assume a more important role in national and international policy analysis.

Larry R. Ford, 1982, p. 133. Planning, after all, is one of the youngest and least coherent disciplines. If we must consistently borrow from it, then we have once more become followers or integrators with little confidence in our own tradition and literature.

Nicholas Helburn, 1982, p. 447. But the major contribution geography makes to our lives is aesthetic: the understanding of what is going on around us, the increased joy that comes from that understanding.

Richard Morrill, 1983, p. 2. . . . while our students have been remarkably successful at finding and excelling in a wide variety of jobs, beyond academia, they are too rarely identified as geographers. We must, then, make clear to the outside world just what is the nature and value of geography, why geography is essential and central to the mission of education at all levels, and why the profession or occupation of geographers should become increasingly recognized.

Shlomo Hasson, 1984, p. 15. If humanistic geography seeks to enhance humanism in geography, it should not confine itself to the sphere of science alone but must enter the realm of relation, thus becoming a socially-committed undertaking. In other words, it should transcend the sphere of gaining verifiable knowledge and start dealing with questions of morality and faith.

David Harvey, 1984, p. 1. Geography is too important to be left to geographers. But it is far too important to be left to generals, politicians, and corporate chiefs.

John R. Borchert, 1985, p. 1. Geographic knowledge is important first, to help an individual understand his or her relationship to the natural and human environments and, second, to help that individual think about what he or she can do to control or adapt to changes in those environments. Because both control and adaptation require not only individual action but also collective action, they demand both individual and public policies. In society, therefore, individual policies, public policies, and geographic knowledge are interacting parts of one system, the purpose of which is to get things done intelligently.

Gilbert F. White, 1985, p. 15. Research intended to serve public policy may begin and end with the expectation that findings, in due time, will be placed in the hands of those charged with policy making. Or it may proceed from problem definition to the drafting of options in close collaboration with policy makers. I favor the latter, interactive process.

Thomas J. Wilbanks, 1985, p. 7. How, as a profession with substantially less than 10,000 people, have we so often come to have a significant impact whenever our best people have turned their attention to national policy issues?

## POLITICAL GEOGRAPHY

Frederick Jackson Turner, 1914, p. 591. There is a geography of public opinion.

Friedrich Ratzel, 1923, p. 270. Similar to the struggle for life, the basic aim of which is to gain space, the struggles of peoples are almost always struggles for the same object. In modern history the reward of victory always was—or was meant to be—a gain of territory.

Helen M. Strong, 1930, p. 211. A geographic background is essential for an interpretation and understanding of the world news coming to us daily in the evening paper, and for an appreciation and an intelligent interpretation of the great trend in modern political developments.

Stephen S. Visher, 1934, p. 72. A consideration of the various types of boundaries from these various points of view leads to the conclusion that the best kind of international boundary today is the kind that has generally been considered to be the poorest, to wit, the arbitrary line located by agreement and artificially demarked. If it runs through sparsely settled areas which have little human interest, such a line is the best kind, because it is most easily demarked, the most definite, and the most permanent. If, on the other hand, it runs through densely peopled regions it is also the best as the most precise and most favorable to a free exchange of goods and ideas. This ready exchange of goods and ideas works toward making the people of both sides of the line similar in welfare and ideals. In so far as this is accomplished, international friction is decreased, and the prospective permanence of the boundary is increased.

Otis W. Freeman, 1936, p. 90. There are always two sides in international controversies and the problems of a nation must be understood before solutions can be offered. A sympathetic understanding of the special racial, economic and social problems of peoples and countries is a most important thing for all students to secure from geography classes.

Owen Lattimore, 1937, p. 549. Even for an industrialized society, imperialism has its point of diminishing returns; the range of action made possible by accumulation of resources and social organization for the purpose of subordinating other societies, different in organization, reaches in time a line that is likely to be vague but that cannot be crossed without overreaching the range of the empire, converting accretion toward the center into a drag away from the center.

Erwin Griswold, 1939, p. 353. To put the forty-eight states into an area of about three million square miles has required the defining and establishment of more than forty thousand miles of boundary lines.

Mark Jefferson, 1940, pp. 60–61. There are striking fringes of German speech around Bismarck's Reich. East of the Reich there is a vast area strewn with islands of German speech . . . —something that leaps to the eye and that other countries lack. These islands of German culture extend across Russia to the Volga. The totality of regions of German speech, German speaking people and German landscape makes up Deutschland. Is Deutschland a country? Was the Reich to which Hitler succeeded in 1934 a country? Is the Reich of November, 1939, a country? Which is Germany? Is there a Germany?

A country is a group of people with feeling of common nationality in their hearts, who accept one government and live on territory that they exploit. The people make a nation and become a country when taken with their territory, but the territory is not their country.

Harold S. Kemp, 1940, p. 61. ("Mussolini 'Geographer.' ") Mussolini, politician, economist and psychologist, is also a geographer. At home he has appraised the Italian resources and found them wanting. At the same time he has . . . justified his vision of a new nation in a renewed country. Abroad he has trespassed only where Italy had much to gain and little to lose. His choice of those environments has not been actuated wholly by greed or vulnerability, but has been so made that, however other powers may have protested, he was assured from the start that they would not substitute violent action for diplomatic usage.

With few exceptions a geographer must applaud Mussolini's reasoning. The man is thus unique among contemporary dictators.

Harold S. Kemp, 1940, p. 140. ("Mussolini: Italy's Geographer-in-Chief.") Like every other geographer, Mussolini loves maps. They figure largely in his propaganda, are used as murals, are featured in the expositions which are now a yearly feature. The most notable propaganda maps—a set of four—carry no printing other than four dates, but there can be no doubting their effectiveness. They portray four periods in the history of the Roman Empire: the meager beginnings in Italy; the expansion into more remote lands; the ultimate greatest extent under Trajan; and the fourth map shows the present empire. No mention is made of the sterility of most of the African possessions and the staring Italians are dazzled by the area of the Empire in comparison with the peninsula alone. The Dictator-Geographer knows that the most ignorant native, looking first at one map and then at the other, is struck by the fact that much of Europe—most of it, indeed—once belonged to the Empire and therefore should belong to it now. Without even a caption these maps are working every minute of the day, and every day. When the Geographer remarks now, however casually, "we really should have this or that territory," he does not have to argue the point. The maps have done that job for him.

Wallace W. Atwood, 1940, p. 342. ("Geography and Great Human Dramas.") If we would have lasting peace in the world, we must break down the barriers of ignorance and establish friendship and cordial relationships between the people of the different countries. We must develop in our people an intelligent as well as sympathetic understanding of the problems in other lands. We must understand the human dramas in progress in the various habitats on this earth depend on education. There is plenty of land for all. There is an abundance of fuel and of mineral wealth in the ground. Our great problem is to learn how to live together and exchange the surplus products of different regions.

Ellsworth Huntington, 1942, p. 1. Geography is changing rapidly. This does not mean that Hitler is altering the political boundaries of Europe. In the long view that is only a minor incident. It means that the science of geography is still evolving. I hope that in the near future it will change in at least four chief ways. First, geography must swing back to a more perfectly rounded type in which man and nature are always considered together, each influencing the other. Second, we need a more clear-cut emphasis upon the idea that distribution upon the earth's surface—the kind of facts shown by maps—is the central theme of geography. Third, the geography of the future will, I believe, ascribe quite as much importance to human differences as to natural resources. In other words, we shall face and ultimately solve, the problem of how far man's physical and psychological status, as well as his cultural status, varies from one part of the world to another in direct or indirect response to the geographical surroundings. Fourth, political geography will come into its own. It will be taught as widely as economic geography, and historians will really utilize it instead of merely paying it lip service.

George Kiss, 1942, p. 633. The Hegelian belief in evolution and in Germany's destiny as a conquering nation and the Kantian doctrine preaching the concentration of all the energies of the nation to one precise end and gearing the nation for one ultimate purpose characterize alike the Germany of the Kaiser and the Germany of the Third Reich.

Derwent Whittlesey, 1942, p. 142. The framework of geopolitics was erected by geographers about 1900. Shortly before the First World War geographers and political scientists brought the earlier work into focus. The war relegated the subject to obscurity; the peace brought it to the fore. With the rise of the Nazi power German geography was merged in geopolitics.

Derwent Whittlesey, ibid., p. 143. Geopolitics has become an engine of political society. It accumulates information for the service of the state, and it functions through propaganda as an arm of government. Originally rooted in geography, geopolitics has its political aspects as needed, because of the large part it plays in the thinking of the German people.

Isaiah Bowman, 1942, p. 646. Geopolitics presents a distorted view of the historical, political, and geographical relations of the world and its parts. It identifies no universal force or process like gravity. It is relative to the state to which it is applied. At least so say its advocates. Its arguments as developed in Germany are only made up to suit the case for German aggression. It contains, therefore, a poisonous self-destroying principle: when international interests conflict or overlap *might* alone shall decide the issue.

Derwent Whittlesey, 1943, p. 97. (''Geopolitics.'') German geopolitics . . . studies the natural environment with a single objective—the improvement of the position of the state with relation to the world total of natural conditions and resources. The state guided by geopolitics disregards conservation in its planned utilization of resources. It knowingly practices ''robber politics,'' akin to the ''robber economy'' of unrestrained and unplanned capitalistic society.

J. Russell Smith, 1943, p. 161. If our intelligence is any greater than that of the sheep or the cow we will strive for international organization that is equipped for:

1. The removal of international tensions before they make explosions.
2. The treatment of any war as we treat smallpox and rabies.

John K. Wright, 1944, p. 190. Political geography is perhaps the most ''human'' phase of geography, since it deals so largely with the strengths, weaknesses, and ambitions of men.

John K. Wright, ibid., pp. 191–192. Geopolitics, I take it, is the geographical study of the contacts of a specific political group, usually a nation, when such study is pursued for the purpose of promoting the interests of that group.

John K. Wright, ibid., p. 192. Military action . . . is governmental political action carried to its furthest extreme, and hence military geography may properly be deemed a highly specialized branch of political geography.

John K. Wright, ibid., p. 194. Political groups are to political geography what landforms are to geomorphology.

Sidman P. Poole, 1944, p. 1. Geopolitik, or geopolitics, has swept the intellectual circles of America in the past few years with its chilling blast of infallibility and invincibility. This ''made in Germany'' brand of thought has inspired some

American geographers to write erudite books on its history, its tenets and influence, while others have been moved to present maps of our future world, beautifully and systematically colored, showing precisely how Germany, Russia, Japan, and other powers must, by the omnipotence of natural laws, swallow up Switzerland and Holland and even China. In many circles there has spread an uneasy feeling that whether we like these arrangements or not, this is inevitable; this is Fate. Geopolitik is simply the revealing of laws inherent in natural phenomena like unto gravity or chemical affinity. This new science spreads before mankind the blueprint that charts its destiny. Let it be stated dogmatically . . . that whatever else geopolitics may be, it is not a logical reasoned science; the systematic presentation of a body of natural laws. It is neither geography nor is it political science as we in America understand these terms. And Geopolitik bears the same relationship to political geography as does astrology to astronomy. The only difference is that astrology merely deludes a few individuals while Geopolitik drenches whole nations in blood. A significant difference.

Preston E. James, 1947, p. 221. The strong isolationist sentiment which kept the United States from assuming its full international responsibilities in the period between the two world wars was, in itself, a form of geographical illiteracy.

C. Troll, 1949, p. 128. Geopolitics, to be sure, has never been acknowledged as a geographic discipline and in Germany has been explicitly distinguished from political geography. But it was an offspring, and finally an increasingly degenerated offspring, of geography.

Anonymous, 1951, p. 234. It is necessary for us to know, not only the contiguous territories of kingdoms and empires, but also their possessions and dependencies in various parts of earth. The treaties between governments, which in modern times are among the most important events recorded in history, are scarcely intelligible to any one ignorant of geography.

George H. T. Kimble, 1951, p. 54. If we want world peace that will last, and a world community (and on purely geographical grounds, the world is small enough, and knit closely enough, to make such a community possible), we need the inspiration and support of religious faith—the kind of faith that constrains men to pass their time not being ministered unto, but ministering, and giving their lives to the service of their fellows. Fortified by such a faith, I believe, ladies and gentlemen, that we geographers could yet live to see the day when, in the words of Tennyson,

> They that dwell apart shall know each other,
> and understand the utterance of a brother,
> in every tongue and tone.

Richard Hartshorne, 1953, p. 393. And because no region or area in this one world can be dismissed as of no significance for survival and progress of all other areas, this unique position of the United States forces upon it a responsibility, for its own future as well as for that of the world as a whole, to concern itself for the fate of countries located in almost every part of the world.

Emrys Jones, 1956, p. 374. A political poll of the percentage of people in favor of certain parties will not in any way affect the results at a subsequent election; the poll has merely predicted, on a sample basis, the probable way in which people would vote if there were an election.

David H. Reeher, 1958, p. 8. Should Intercontinental Ballistic Missiles, our primary strategic weapon system of the near future, be based outside the United States?

W. A. Douglas Jackson, 1958, p. 178. Has there ever been a period in history other than the present when man has been so troubled about his behavior and the complex of motivations which drive him simultaneously toward great achievements and self-destruction?

James M. Hunter, 1958, p. 271. Diplomacy has a geographic foundation. In each major action in the realm of diplomacy of all states, geographic factors are found embedded within the problem. Recognition, understanding, and use of these factors is advantageous since they are more permanent than are the political conditions of the problem. Hence, they give the diplomat a foundation on which to build the policy of the function of the state. Failure to recognize the geographic factors may result in weak, faulty policy that will generally defeat the effective function of the state.

Chiao-Min Hsieh, 1959, p. 186. Since the Soviet Union's yesterday is Communist China's today, geography, like other sciences in Communist China, has had to be reoriented to follow the Soviet pattern.

Peirce F. Lewis, 1959, p. 195. Politics differ from place to place, and the differences can be shown on maps of election returns. Political attitudes, however, are related to certain population characteristics, which are also mappable. A comparison of maps, therefore, should reveal characteristics of group voting, and the present study is designed to test how effectively such comparisons can be made. Such a test is desirable because students of American voting behavior have often used maps as illustrations or as means of recording data, but seldom as analytic tools.

Ladis K. D. Kristof, 1959, p. 281. The nature of frontiers differs greatly from the nature of boundaries. Frontiers are a characteristic of rudimentary socio-political relations; relations marked by rebelliousness, lawlessness, and/or absence of laws. The presence of boundaries is a sign that the political community has reached a relative degree of maturity and orderliness, the stage of law-abidance.

Geoffrey J. Martin, 1959, p. 443. Geopolitics has served to focus attention upon political geography, for it has absorbed the data and framework supplied by the inquiry of the political geographer. This body of planned and calculated thought, which uses the broader frame of reference and mass of data afforded by political geography, has presented itself at one time or another and with varying degrees of success to the governments of most centuries of the world. Historically, geopolitics arose with the nation-state system and has ever since functioned as an instrument of government: the recent development of German *Geopolitik* as the political conscience of the state has further insured geopolitics a position in contemporary state government.

Geoffrey J. Martin, ibid. Geopolitics is essentially a body of thought developed in a given territory which seeks the maximization of its own ends. The geopolitician sees all other groups thru the spectacles of national interest, often becoming selfish to the point of greed, lust, and violence. The core of this discipline is power; the quest for power provides the guide to method.

James M. Hunter, 1960, p. 327. The organic theory of state growth and development from the biological viewpoint has been proved false and unusable in geographic methodology. When this theory was utilized in actual implementation, as the motivating idea for the internal and external division of state policy, it led to disastrous results.

Preston E. James, 1963, p. 97. Democracy and autocracy are the words that stand for two incompatible systems of living. We are in the midst of a world-wide conflict between the supporters of these systems for the control of men's minds, and if a shooting war intervenes this will only postpone the ultimate decision between them. In the face of this situation if young people are to be prepared to perform the duties of responsible citizenship and to face the changing world without fear they must gain an understanding not only of the historical processes involved, but also of the differences from place to place and country to country on our contemporary earth. As never before they need the perspectives of both time and place. If geographers are to move forward with the opportunity now at hand, and if they are to meet the challenge of this age of revolution, then they must so structure the concepts and content of their field that its relevance to the modern world is immediately apparent.

John K. Wright, 1963, p. 1. The German geographers who developed *Geopolitik* at the behest of their imperialistic masters manifested a certain rapacious ferocity, but, like bulls, they had rings in their noses.

Karl Marcus Kriesel, 1968, p. 557. It has been generally accepted that the field of political geography includes the study of precisely what it is about a state's environment which has merited its political demarcation. Some political and geographical thinkers have held the opinion that it was the character of the physical environment which influenced the size of a state and the form of its government. The shades of meaning of the word "influence" have ranged from the severe "determine" to the mild "suggest."

Norton Ginsburg, 1973, p. 15. The devising of a locational strategy for national development ought to be a problem made to order for the geographer. Yet, the geographical literature on this important subject is remarkably sparse.

Norton Ginsburg, ibid., p. 18. One cannot speak of development in Asia without referring to the Japanese experience. Although one cannot expect that what happened in Japan in the last hundred years will be repeated elsewhere, if only because other Asians are not Japanese and because the contextual environment was so very different than it is now, the Japanese approach to planning even today may have some valuable lessons for development in the presently poor countries.

Clark Akatiff, 1974, p. 26. The weekend of October 19–21, 1967, marked a major turning point in the development of militant antigovernment protest in the United States. This was the weekend when some 400,000 people converged on Washington, D.C., in an attempt to "Confront the Warmakers." I . . . believe that geographic analysis of distribution, dispersals, flows, and environmental perception provide powerful analytic tools for the general understanding of these movements.

Julian Wolpert, 1976, p. 10. Institutions, once created, persevere to remain, to grow more prominent, and to take on additional functions.

David Harvey, 1979, p. 381. The building hides its secrets in sepulchral silence. Only the living cognizant of this history, who understand the principles of those who struggled for and against the "embellishment" of that spot, can truly disinter the mysteries that lie entombed there and thereby rescue that rich experience from the deathly silence of the tomb and transform it into the noisy beginnings of the cradle.

All history is, after all, the history of class struggle.

Ross Terrill, 1980, p. 240. Mao knows much about the outside world. . . . No other world leader of midcentury—not even de Gaulle—read for himself and wrote his own talks to the extent Mao did. History and geography were the twin pillars of his foreign knowledge.

Hendrik-Jan A. Reitsma, 1982, p. 126. Dependency theory leads one to focus exclusively on the relations of Third World countries with advanced capitalist countries. Scarcely any attention then is paid to the socialist states because they are, according to the trichotomous classification, neither developed nor underdeveloped. The upshot of this is that the discussion and analysis of exploitative relations, deteriorating terms of trade, self-serving aid programs, and so forth, is confined to the capitalist system, as if there were no dominance-dependence relations among socialist countries or between advanced socialist countries on the one hand and less-advanced nonsocialist countries on the other. Thus the socialist developed countries go scot-free, even though there is ample evidence that their trade and aid practices vis-à-vis Third World countries differ little from those of the Western core countries.

David Harvey, 1984, p. 4. The geographical literature can express hopes and aspirations as well as fears, can seek universal understandings based on mutual respect and concern, and can articulate the basis for human cooperation in a world marked by human diversity. It can become the vehicle to express utopian visions and practical plans for the creation of alternative geographies.

David Harvey, ibid., p. 5. The failing credibility of positivism in the late 1960s opened the way to attempts to create a more directly radical or Marxist tradition. Geographers were faced with a peculiar mix of advantages and disadvantages. Old-style geography—global, synthetic, and dealing with ways of life and social reproduction in different natural and social milieus—lent itself easily to historical materialist approaches, but was dominated by establishment thinkers attached to ideology of empire or actively engaged in the service of national interests.

## MAPPING SCIENCE

E. C. Bentley, undated, p. 42.

Geography is about maps,
But Biography is about chaps.

Douglas Wilson Johnson, 1910, p. 156. If you wish an example of the mental confusion which arises from a bad memory map, ask a person whose mental picture of Africa has the Equator crossing the Sahara Desert (there are multitudes of such persons), to locate the Equatorial Forests of Africa.

C. F. Close, 1910, p. 750. I wish it were possible to prophesy smooth things about Ceylong. From our special point of view the situation leaves much to be desired. There is not yet published a single topographical map, and the topographical surveys are progressing at a rate which, under favorable conditions, may result in the maps being completed in the year 1970.

Cyrus Adams, 1912, p. 194. Map-making is very old, and has been practised by the most primitive peoples for many ages. Rude scratches on many rocks in South America are now interpreted as maps. East Greenland natives carve maps out of wood; American Indians make map-sketches on birch and other barks; the Marshall Islanders charted the sailing routes along their coasts long before they knew of the white man; the desert nomad sketches maps in the sand to illustrate his wanderings, and nearly every primitive tribe to-day makes maps to show routes to hunting-grounds, animal paths, fisheries, fords, etc. They know as well as we do that maps are practically a human necessity; but we know further that a good map often places before our eyes an amount of accurate geographical information that might take many months to dig out of books.

Martha Krug Gentha, 1913, p. 33. No single cartographer can therefore produce equally good work in every field of cartography, and in spite of occasional utterances to the contrary, a certain diversity of styles is always found even within one and the same country or one and the same institution.

Henry Wilson, 1918, p. 196. There is a geography of thought, a geography of spirit, a geography of psychology, of racial influence, a superphysical geography—in fine a geography. We want maps of mind, showing the thought and cultural currents, idea drifts, spiritual isobars, contours of artistic altitudes.

Louis Karpinski, 1923, p. 212. Any serious study of the development of a modern map reveals then that in part we are truly the heirs of all the ages. The maps come to us as a part of our heritage of science. The map is the achievement of countless scientists of ages past, of all nationalities seeking to record their comprehension of the universe about them. The map of the world as we have it today is a symbol of the unity of the sciences; the map is a symbol of the progress of science as the product of scholars of every land and of every age.

W. L. G. Joerg, 1923, p. 211. [Airplane photography] is especially valuable in the study of American cities.

J. Paul Goode, 1924, p. 39. Mercator's projection was devised as an aid to navigation. For this purpose it has no superior. It was never intended as a world map for distribution, and other geographic uses.

J. Paul Goode, 1927, p. 1. The map is a written language expressed in a system of shorthand. In its earlier forms it was usually pictographic, and this one trait has clung to it throughout the ages. So nearly as we may tell from the study of primitive cultures, the making of a map is as old as any other form of picture writing. In fact it has been considered close to a primitive instinct. And as a form of graphic expression it was practiced in every geographic realm from the most ancient times.

Guy Elliott Mitchell, 1929, p. 382. The ancient geographers invested unknown areas on their maps with fabulous and wondrous inhabitants—giants, strange and terrible animals the like of which man has never seen, vast and impenetrable regions, stupendous mountain ranges, and other features indicating in reality the scope of the imagination of the map makers.

Erwin J. Raisz, 1931, p. 297. There is one problem in cartography which has not yet been solved to general satisfaction and which is a very important one: the depiction of the scenery of large areas on small-scale maps.

Robert E. Hartz, 1931, p. 339. Maps were placed in glass-covered rectangular boxes, fastened at convenient points in cockpits, with controls at the sides, making it possible to rotate the maps forward or backward according to the immediate territory over which the pilots were flying. An operations order for flight included the time, place, and elevation of rendezvous. From this point the formation would fly from point to point during the required time, in an attempt to carry out the objective assigned. One can readily see the importance of an understanding of map reading in such assignments to aviators. Each pilot was obliged not only to watch his plane, his place in formation, the elevation, time, gas and oil supply, his machine guns, and the enemy, but he was to fly over the prescribed territory called for by checking what he saw below with the map before him.

Guy-Harold Smith, 1932, p. 76. The hobby of stamp collecting should be of more than casual interest to geographers. The map stamps in particular have sufficient geographic character to permit serious study.

Margaret Irene Fead, 1932, p. 56. Two types of city maps have been developed, the perspective and the horizontal, of which the latter is the earlier.

Floyd A. Stilgenbauer, 1932, p. 77. The function of a population map is to give a correct impression of the areal distribution, density, agglomeration, and group relationship of people. A population map is indispensable in the study of modern geography.

Joseph Hyde Pratt, 1932, p. 233. With the realization of the ancient Greeks more than twenty centuries ago that the earth has a spheroidal surface, the logical position for a datum line from which north and south measurements of latitude might be made was clearly established. However, the nature of things provided no such obvious location for a primary line to which east and west determinations of longitude might be referred. As a consequence, each map maker chose what seemed to him an appropriate point through which to draw his prime or reference meridian. Eratosthenes, perhaps the first scientific cartographer, based his zero line of longitude on Alexandria; Hipparchus, on the island of Rhodes; Ptolemy, on the westernmost known bit of land, the Canary Islands.

Mark Jefferson, 1932, p. 201. Like all students of things geographic, I have always had to put data on maps of the world, and I have found all world-maps too small. I always want the continents larger, and their shape is not like continent shapes on the globe. I want my continents of truer, better shape.

Mark Jefferson, ibid., p. 204. There is so much ocean in the world! We cannot afford to include so much water in our maps when they are concerned with land affairs only, and these are of course most affairs since man is in the main terrestrial in his habits.

Margaret Irene Fead, 1933, p. 441. The first known city maps are those from Mesopotamia. A crude and simple map of Babylon, done on a clay tablet, is an attempt to show the areal spread of a city and is of great antiquity [dating from about the middle of the 7th century B.C.].

William Bowie, 1933, p. 42. The making of accurate charts and maps is a matter of great importance to geography. There are very few geographical matters not based on charting and mapping information.

Erwin Raisz, 1934, p. 292. The idea of the statistical cartogram occurred to the author when he had occasion to prepare maps of the United States showing the distribution of various economic units, such as steel factories, textile mills, power plants, banks, etc.

Wellington D. Jones, 1934, p. 105. Of the various forms of record employed by geographers—written statements, tables of data, diagrams and graphs of many kinds, photographs, and maps—the map ranks first as a device for showing the areal extent and pattern of distribution of phenomena.

Robert Burnett Hall, 1935, p. 126. During the period of environmentalism, American geography drifted far from the morphological approach to the features of the natural environment, and largely forgot its own child and best servant the map.

Robert Burnett Hall, ibid., p. 127. [A] critic writes: "I am becoming more and more dissatisfied with the regional studies submitted as partial fulfillment for the Doctor's degree. At best, they are but highly superficial examinations of some fragment or other of the earth's surface." Such criticisms are frequent and have been voiced both from within and without the field of geography. We have strangely failed to develop the central discipline of our science which lies in the technique of map making and the interpretation of the map.

Erwin Raisz, 1936, p. 8. The rectangular statistical cartogram is a device for visualizing statistical facts. Its principle is, that each region involved in the study is represented by a rectangle, the area of which is proportional to the value which this region represents. The rectangles are put approximately in the same position as they are on the map.

Erwin Raisz, ibid. Since the first statistical cartograms of the United States were published in 1934, considerable interest has been shown by teachers in this device to use it as a tool in teaching geography.

Joseph A. Russell, 1939, p. 91. Vertical aerial photographs contain a wealth of geographic inventory data, some of which is directly interpretable from the photographs, and some of which must be translated into more understandable terms before it can be used.

Richard Hartshorne, 1939, p. 425. So important, indeed, is the use of maps in geographic work, that, without wishing to propose any new law, it seems fair to suggest to the geographer a ready rule of thumb to test the geographic quality of any study he is making: if his problem cannot be studied fundamentally by maps—usually by a comparison of several maps—then it is questionable whether or not it is within the field of geography.

Claude H. Birdseye, 1940, p. 1. The greatest advance of the past twenty-five years in map making is undoubtedly the application of aerial photography to the preparation of planimetric and topographic contour maps.

J. E. Spencer, 1941, p. 79. The fascination of maps derives not a little from the language spoken by their cover of place names. Many of these have grown out of the everyday life and experiences of the inhabitants. Mixed in among these lowly native names may be regional names supplied by explorers, geographic boards, or planning commissions, by historical events or changing political boundaries. There are the recurring names that evidence migration of a people or a culture; and there are the place names taken by a people of one culture from the people of another, such as the Indian and Spanish names throughout much of North America. World maps prepared for use within a given country misrepresent in some degree the place names of countries of other speech. When different alphabets are involved, the misrepresentation is magnified; and when a different system of language is used, as in China, the distortion puts the cultural and geographic realities completely beyond the grasp of the layman.

E. H. G. Dobby, 1942, p. 211. The study of settlement pattern in equatorial environments is ordinarily prohibited by the lack of adequate topographic maps.

John K. Wright, 1942, p. 527. Like bombers and submarines, maps are indispensable instruments of war.

John K. Wright, ibid. Maps help to form public opinion and build public morale.

John K. Wright, ibid., p. 529. Beware of maps prepared to substantiate a pet theory! There is a well known type of reasoning that begins with a theory, gathers statistics and other data that seem to support it, makes a map on the basis of the statistics, and finally "proves" the theory by reference to the map. The dishonesty in such a procedure may be unconscious, but there is a large use of maps in propaganda with a view to conscious and deliberate deception in the service of special interests. The relative areas of different regions as disclosed on maps in railroad timetables are usually deliberately distorted so as to show particular railroad systems to best advantage. More subtle and dangerous is the type of deception found on maps designed for propaganda purposes—maps on which facts are played up or played down, omitted or invented, for nationalistic ends.

John K. Wright, ibid., p. 543. If map makers are human, so too are map users. Like carpenter's tools, maps should not be misused.

F. J. Marschner, 1943, p. 199. Maps constitute one of the principal depositories of knowledge pertaining to conditions on the earth's surface.

Louis O. Quam, 1943, p. 21. The map, the most distinctive tool of geography, has become an important propagandic weapon in psychological warfare. German geo-politicians discovered that the public has great faith in the authenticity of data depicted on maps. They found, therefore, that one of the best methods of presenting an idea to the people is to represent it on a map. Careful attention was given to methods by which maps could be made to create a desired false impression from otherwise accurate data. In this way propaganda maps are designed to produce impressions rather than to reveal information. The map becomes a psychological force instead of a scientific tool.

Russell H. Fifield, 1944, p. 297. Only 5,000 years and 4,000 miles separate the pioneers of Western civilization in the river valleys of the Middle East from the aerial pilots of the future in the Mediterranean or the Arctic.

Henry J. Warman, 1944, p. 304. Persons trained in cartography and graphics are in great demand during this war. Magazines, newspapers and business concerns now are using and will continue to use cartographic displays and graphic devices to facilitate the public's grasp of news and advertisements at a glance. This field of geography, the use of the tools, expands obviously into numerous war and peacetime positions for the boys and girls who show ability along this line once the opportunity presents itself for them to develop it.

Kirk Bryan, 1944, p. 184. The representation of data on maps is the heart of the geographic method in all fields.

Kirk Bryan, ibid. The geographer must understand not only the general methods of map-making or cartography, but much of the detail so that he may understand the limitations of method and the degree of precision attained.

Kirk Bryan, ibid., p. 185. The use of maps as vehicles of propaganda has been advocated recently by the "geopolitical" pseudo-geographers of Germany. Obviously they are unfamiliar with the efforts of the American advertiser. The brochures of our chambers of commerce, the familiar railroad time-tables, the advertising pages of our magazines all contain maps used as propaganda. Distortions, overemphasis, violent contrasts in shade and color and other devices are the common aids to creating the mental effect intended. Our youthful geographer might well study these skillful artifices of the advertiser. No one would advocate that the young geographer should indulge in the cartographic extravaganzas of the advertiser. He might, however, learn simplification and emphasis, qualities that would vastly improve our current geographic literature.

Richard Joel Russell, 1945, p. 305. What seems to be a rather widespread belief that the airplane has revolutionized geography and cartography is probably the most glaring admission of ignorance about these subjects in vogue today. Will thirst for fads lead us to the superredundancy, "A global geography of the world?"

Erwin Raisz, 1946, p. 347. With the end of the war it was expected that the unprecedented war-production of maps would fall off to peacetime level. This, however, was not the case. Never again does the United States want to be caught so unprepared with maps as at the beginning of the war, and the great cartographic projects initiated during the war are still being carried out. The personnel of the various map making agencies was reduced perhaps to half of the peak employment, but this is still five to ten times the peacetime level.

Glenn T. Trewartha, 1948, p. 170. The fact that more than 80 per cent of the continental United States has now been photographed from the air immediately suggests the possibility of using aerial photography in the study of farmsteads.

George H. T. Kimble, 1951, p. 45. History does not record the name of the man who first reflected that it was a small world.

Allen K. Philbrick and Harold M. Mayer, 1951, p. 367. The graphic problem in the simultaneous examination of more than one pattern in a single map has always taxed the ingenuity of the cartographer.

Carl O. Sauer, 1952, pp. 101–102. No field of inquiry can be properly defined by any specific means or methods of gaining knowledge. A person may devote himself to the mastery and use of a particular technique, but knowledge is eclectic and grows by using any means that add to understanding. Geography, for example, has a vested interest in what is called by everyone the "geographic method." In simplest terms, this is placing on a map the distribution or variation of anything that is spatially localized or varying. Geographers should be making more use of this method of inspection than they do, for many questions of distribution cannot be studied without it. I should not admit, however, as has been claimed, that all the data that concern geographers are mappable, or that their conclusions are necessarily derived therefrom. Therefore, I cannot agree that geographers can be recognized by their own method. Some can be, some good ones cannot be.

Mildred Danklefsen, 1953, p. 254. Geography lends itself readily to televising.

Fred K. Schaefer, 1953, p. 244. The morphological character of geography finds its expression in its own specific tool, maps and cartographic correlation. Mapping has been called the shorthand of geography.

Clarence L. Vinge, 1953, p. 8. The progress of our science has been seriously handicapped by the custom and tradition of using the written word as the chief vehicle of most geographic communications. . . . Few geographers seem to be willing to commit themselves to the thesis that the map is (or should be) the principal language of geography. Why are we so unwilling to accept this thesis?

Edward L. Ullman, 1953, pp. 56–57. The geographer uses the map as his primary tool. I do not mean principally cartography, nor the surveying, drafting, or reproduction of maps, although here again is a limitation that some social scientists would logically, but mistakenly, place on the geographer's role. I mean the use, interpretation, and imaginative compilation of maps for the purpose of showing spatial interrelations. One might argue that the geographer need know no more of the drafting and reproduction of maps than the novelist need know of the mechanics of printing and setting type. The geographer also would probably take responsibility for the maps in a joint area research project and hire the draftsman, a not unwelcome service, but in itself merely a detail of the geographer's contribution.

J. Wreford Watson, 1953, p. 323. The principal techniques which geography uses are field surveys, the interpretation of aerial photographs and the making and reading of maps.

O. Orland Maxfield, 1953, p. 25. There are many roads which may be traveled in the study of geography. Some of them are broad and smooth, the map having been filled in after careful surveys to show the major attractions, the treacherous stretches, the best routes by which the traveler may achieve his specific goal. But there are others which are mere lanes, seldom frequented ways, poorly charted and beset by many obstacles.

Kirk H. Stone, 1954, p. 318. Recently attention has been refocused on the utility of air photos. Pictures have been taken from fast-moving vehicles at tree-top elevations and from vehicles in earth-girdling orbits. Interpretation of them has been dramatized, and their values as an analytical tool have been publicized.

Carl O. Sauer, 1956, p. 289. If geographers chance to meet where maps are displayed (it scarcely matters what maps) they comment, commend, criticize. Maps break down our inhibitions, stimulate our glands, stir our imagination, loose our tongues. The map speaks across the barrier of language; it is sometimes claimed as the language of geography.

Carl O. Sauer, ibid. Show me a geographer who does not need them [maps] constantly and want them about him, and I shall have my doubts as to whether he has made the right choice of life.

Robert B. Monier and Norman E. Green, 1957, p. 2. Generally speaking there has been little utilization of aerial photography in *human* urban geographic research.

Virginia P. Finley, 1958, p. 261. Two important uses of aerial photographs for basic geographic research are (1) for conducting comprehensive regional surveys and (2) for tracing changes in the landscape over a period of time.

Howard G. Roepke, 1958, p. 11. Much has been written about the storage, care, and indexing of sheet maps. Wall maps have been largely neglected—both in print and by the geography departments using them. They are essential to the effective teaching of geography, yet little thought is usually given to achieving the maximum utility from a collection.

G. D. Taylor and C. M. Matheson, 1958, p. 13. Geographers are often faced with the problem of preparing reports that require many maps. The preparation of these is usually time consuming and often delays the presentation of the report. We have experimented with the use of a photo copy machine in an effort to reduce the time involved in map preparation while not reducing the quality of the map. These experiments have convinced us that these machines are of great value in reproducing black and white maps of report size.

Arch C. Gerlack, 1958, p. 262. Linguistic maps present a real challenge to the cartographic profession. To improve their generally poor quality and ineffective design is not an easy task, but it is a worthy objective. Language maps are well above the median for the use curve of special subject maps and they serve a wide variety of disciplines.

Benjamin F. Richason, Jr., 1959, p. 317. In geography, one difficulty frequently encountered in presenting material is the inability of the student to see and to comprehend much of the earth's surface at any one time. Maps, charts, diagrams, and photographs which have the effect of reducing large areas of the land to observable size have been used by earth-bound scientists as teaching devices. Now, an effective and impressive method of accomplishing this is observation classes from an airplane.

John E. Dornback, 1959, p. 179. All normal individuals possess the ability to form schematic mental maps of routes to and from familiar places. It is also possible to form mental maps through study of places never visited. Contrary to the very refined mental maps of the geographer, it is not exceptional to find intellectual individuals with such poor mental maps that they have little concept of the relative position of continents.

William D. Pattison, 1960, p. 4. Although photography's alliance with geography is perhaps stronger than ever before, the relationship has in recent years suffered one distinct loss. This has occurred with the passing from the geography classroom of the traditional, large (3 1/4 x 4 inch) lantern slides, with their photographs printed directly upon glass. Faced with competition from cheaper, lighter, and smaller slides of mounted film, the old glass slides have nearly ceased to be produced, and such collections as have survived in geography departments are to be found most often in dark corners and back rooms.

Marvin Gast, 1962, p. 333. The versatility of the overhead projector is unmatched by any other single type of visual aid equipment. This year, for the first time in the history of the Association of American Geographers, the overhead projector will be used to support a paper presented at the annual meeting. It will doubtless be used frequently in the future.

Andrew H. Clark, 1962, p. 236. It is now accepted doctrine among us that a cartographer who is not also a well-trained geographer suffers a crippling handicap in the performance of his own tasks.

James P. Latham, 1962, p. 346. The complex geographic patterns of phenomena now recorded or transmitted by aerial photography, image sensors, or other environmental sensors provide distribution data that require rapid and rigorous methods of quantitative analysis.

George F. Jenks, 1963, p. 599. Geographers have long been proud of the fact that they excel in the understanding and use of maps.

W. R. Tobler, 1966, p. 351. The study of ancient maps provides one of the fascinating aspects of historical geography. Modern theories regarding the ancients' perception of the world also may require consideration of the map projection employed for maps.

John E. Estes, 1966, p. 673. Aerial photography has long been recognized and accepted as an important tool in geographic research. Although aerial photography is no substitute for thorough field work, a working knowledge of aerial photographic interpretation techniques is of immeasurable value to the geographer in many types of field investigations.

Charles E. Olson, Jr., 1967, p. 382. Infrared technology has advanced so rapidly that airborne sensing systems can now generate imagery of such high quality that it resembles conventional aerial photography.

Preston E. James, 1967, p. 3. . . . it [space] is a curved surface, and very much larger than the observer. It has never been possible for the individual geographer to observe closely any large part of the earth. Even today with the new techniques of remote sensing, the individual geographer will not be able to see for himself any large part of the earth during a normal lifetime. The geographers are forced, therefore, to gather much of their knowledge about the earth at second hand.

Meredith F. Burrill, 1968, p. 7. Today, from automated instrumental observation from high flying aircraft and spacecraft, remote sensing as we call it, we may get more reliable information, in much greater quantities, without time wasted going out after it. This could be revolutionary. But it might take a while, and the linguistic problems involved in developing effective forms of communication are going to be significant and serious.

John R. Borchert, 1968, p. 371. Remote sensing promises to contribute much to the evolution of geographic science.

Raymond E. Crist, 1969, p. 306. The use of inadequate or insufficient data may lead one to bridge gaps, or even wide, inductive chasms, by making highly subjective inferences, from which in turn misleading assumptions or even conclusions might be drawn. And when conclusions are embodied in beautifully wrought, symmetrical maps they seem to be so solid and basic, so firmly fixed in a matrix of immutable facts, as to give the lie to their subjective content.

Kirk H. Stone, 1970, p. 6. It seems paradoxical that in modern times we should become larger scaled in our studies. While transportation media have extended and speeded up, and often remained the same in passenger-mile cost, our published results have declined in scope. Rarely does one find an article in our journals covering the mapping of a continent, much less of the world. Most world maps are in elementary textbooks, not advanced articles or monographs.

Arthur H. Robinson, 1979, p. 97. In the modern history of cartography, it is a fact that periods of war have generally led to developments in mapmaking.

Phillip C. Muehrcke, 1981, p. 397. Academic relations among geographers, maps, and cartographers have deteriorated dramatically over the past several decades. Even though annual map production is at an all-time high, the proportion of geography professors who regularly use maps in their teaching and research has declined.

Phillip C. Muehrcke, ibid., p. 404. Geographers who avoid maps needlessly limit their ability to conduct geographical research and communicate geographical information. Words and numbers alone are not sufficient means of expression in the environmental sciences, if in any field. In geography, cartographic expression is crucial. As environmental problems become more difficult to solve, jobs will increasingly go to those who are able to merge words, numbers, and cartographic methods into a timely and integrated approach to geographical study and exposition.

Richard Morrill, 1983, pp. 6–7. . . . space—in the geometric sense of separation and relative location, in the environmental sense of the unique identity of places—is the core of geography. It follows that the map, and cartography, are absolutely fundamental parts of the discipline, not only in depicting geographically distributional patterns, networks, flows and structures but also as a key research tool, in communicating geographic and spatial ideas, generating hypotheses about spatial behavior and processes and recognizing structures it is geography's purpose to explain.

Jerome E. Dobson, 1983, p. 135. Recent advances in analytical methods and computer technology have made automation possible for almost every scientific procedure that heretofore was performed manually in geographic research and problem solving.

## FIELD TECHNIQUES

Charles R. Dryer, 1911, pp. 9–10. The landscape is not made up entirely of earth and rock. It is clothed with vegetation, natural and more or less modified by human action, and this living garment of the land is passing through a series of changes parallel with the changes of land material and form. The soil forms the connecting link between the earth crust and its clothing and determines very strictly warp, woof and texture.

Charles R. Dryer, ibid., p. 10. By the harmonious cooperation of geographers and botanists a new science has been added to the conquest of the world. The natural vegetation of a region is the fullest and most suggestive expression, on the one hand, of the physical conditions of structure, relief and climate which prevail there, and on the other hand, of the natural controls exerted upon human activities. Vegetation is the key which unlocks the chain of geographic causes and consequences.

Harlan H. Barrows, 1923, p. 14. Geography has been too much a library subject, and too little a field subject. I hold that the field is the geographer's laboratory.

Wellington D. Jones and V. C. Finch, 1925, p. 148. Sound conclusions in geography, as in any other subject, must be based on facts. A large proportion of the facts needed by the geographer can be obtained only in the field, and since observations of workers in other sciences and of untrained travelers have proved quite inadequate for the geographer it is clear the latter must make and record his own fundamental observations as a basis for description and interpretation. Field maps constitute a vital part of the record of these observations, and a problem of primary importance therefore is the determination of what observed facts shall be mapped and how the mapping shall be done.

D. S. Whittlesey, 1927, p. 72. It is only a little more than ten years since students of Human Geography in the United States began to publish specifications for field study in their subject. For some time public discussions of field technique remained theoretical in tone, and it has been less than two years since results of practical experiments first came from the press.

Wellington D. Jones, 1931, p. 208. At the beginning of our work we formulated as a general objective, the discovery, analysis, depiction, explanation, and appraisal of land occupance and use within the metropolitan area. We agreed that discovery, analysis, and depiction of the facts of land utilization should precede explanation and appraisal of these facts, and that probably the most direct and effective way of acquiring many, if not most, of the basic facts was to make, in the field, a map of land utilization.

Carl O. Sauer, 1941, p. 14. ("Foreword.") Let no one consider that historical geography can be content with what is found in archive and library.

Carl O. Sauer, 1956, p. 296. The principal training of the geographer should come, whenever possible, by doing field work.

Erwin Raisz, 1957, p. 176. Geostenography is a system of geographical notetaking that was developed during the excursions after the Congress in Brazil. It was particularly helpful in a shaky autobus, passing rapidly-changing scenery. The method consists of symbols, letters, numbers, and profiles. For example, 200M is a mature land with 200 meter relief. 50V $U_7$ is a plateau with rivers steeply incised down to 50 meters and with 7/10 upland. $Ba_2P_5$ is a region with 2/10 in bananas, 5/10 in pasture, and the rest is capoeira (cut-over forest). Land profiles are indicated graphically.

Hans Carol, 1960, p. 309. A student receives most of his training indirectly from professors and by literature. He comes in direct contact with geographic reality only in his field training. Here he gets the chance to transfer reality into intellectual concepts; here he observes correlations and practices the first steps of generalization.

Andrew H. Clark, 1962, p. 237. Perhaps the saddest thing about the decline of skill and practice in field observation is not the methodologic danger, grave though that may be, but the denial, thereby, of one of the potentially greatest of a geographer's incidental rewards.

John K. Wright, 1963, pp. 1–2. I have known wild geographers because I have spent most of my life in places that they frequent, as animals do salt licks—notably Cambridge, Massachusetts, and New York City, where Harvard and the American Geographical Society respectively afford delectable accumulations of "salt." I have also run with the wild pack on field trips, and there is no better way of getting to know geographers.

## GEOGRAPHIC EDUCATION

Anonymous, 1813, p. 235. To all classes of people is geography important, but to students it is doubly so. They are supposed, when they leave the University, and go into the world, to have treasured up & made themselves familiar with such knowledge as will be most useful to them in life; and as none is more so than geography, if they are ignorant of that, they will unavoidably be forever exposed to serious mortifications among those of inferior talents and far less acquirements than themselves. To be ignorant of what it is supposed one ought, from the advantages he has had, to know, argues stupidity or culpable idleness, and who is ambitious to be thought a dunce or a lounger?

J. Paul Goode, 1911, p. 156. We defeat our educational purpose by attempting to cram into one elementary course, conditions, commodities, commerce, and countries; and that it will be a most profitable program to divide the field, and to devote the elementary work to the commodity and its travels.

W. M. Davis, 1911, p. 157. ("Commercial Geography.") If school and college courses in commercial geography are intended eventually to contribute to the development of international commerce, they ought to include some anthropology or ethnology; enough, indeed, to make it clear to every student that the manners, customs, beliefs and morals of other peoples than his own deserve sympathetic study and respectful consideration, even if on no other ground than a selfish wish for more profitable business relations as a result. There can be little question that a recognition by our early frontiersmen of Indian ideas regarding right and wrong would have diminished the wretched conflicts that marked our advance inland from the Atlantic coast; but our aggressive frontiersmen were usually self-selected on grounds far different from those of a gentle consideration of the Indian's point of view. There can be no doubt that a better understanding on our part of Mexican formalities and courtesies would lessen the misunderstandings and animosities that are only too frequent on both sides of our southern border, as well as facilitate harmonious commercial intercourse with our next-door neighbors; but the alert and pushing American too often learns little and cares less about Mexican manners before undertaking his business journeys, and he therefore frequently offends those whom he would be glad, even only on the selfish ground of business profit, to please.

Eugene Van Cleef, 1913, p. 235. Geography as taught in the schools today may be classified under two heads namely, Geography of Fact and Geography of Interest. The first kind of Geography is characterized by the old-fashioned "grind"; the second kind by the new-fashioned "lack of effort."

Richard Elwood Dodge, 1916, p. 3. Education is not today what it was twenty-five years ago. No longer are the best methods of teaching a given subject decided by show of hands in a class room or applause at a teacher's institute. A successful worker in education must follow those fundamental principles of his young science, that have been measured quantitatively or tested and proved by other experimental methods. The apprentice system, by which so many of us have come to our life work, cannot now be considered the best preparation for success in teaching any subject.

Richard Elwood Dodge, ibid., p. 6. A student knows his geography well when he knows the location of all the important places and products and can tell why these are where they are; and when he knows all the important geographical facts, as every person should by the time he finishes the geography course.

Richard Elwood Dodge, ibid., p. 13. Geography, which we believe has worth that cannot be challenged, must be approached from a new standpoint. Few realize that there is any other geography than the physical and commercial geography of the last few years, though several reports advocating the development of regional geography, and the humanizing of physical geography have been widely distributed and much discussed in recent years.

W. M. Davis, 1924, p. 199. It is in human geography that we owe the most to European teaching; first . . . , to Guyot; later and in larger degree to Ratzel, whose views on anthropogeography, altho represented in America chiefly by a single one of his pupils, have had great influence among us; later still, but as yet in less degree, to the French school of historical geographers of which Vidal de la Blache was the leading exponent.

Leonard O. Packard, 1924, p. 147. The longer I teach geography the more I find myself giving time and attention to the use of interesting detail.

Albert Perry Brigham, 1924, p. 109. ("Association of American Geographers.") A vast total of scientific work done in America during the nineteenth century, had of necessity been of a geographic nature, though not always recognized as such. During the last decade of the century there was a growing recognition of the fact that in secondary and higher education geography deserved a place of its own. Following European example it was coming to be understood that the science was too inclusive and too important to hover in fragments under the shadow of other subjects.

Eric P. Jackson, 1924, p. 313. Geography as a school subject is particularly well adapted as a correlating center for other subjects in the curriculum. Starting as it does with the powerful influences of our physical environment, such as relief, soil, and climate, geography soon develops intimate and inseparable relationships with many other subjects. This interrelation is to be found not only in the "tool" subjects, such as arithmetic, English and the languages, but to a larger extent perhaps in such "content" subjects as history, economics, literature and science.

G. T. Renner, Jr., 1926, p. 272. ("Geography's Affiliations.") So instead of trying to force a form of geology back into the curriculum under the name of geography would it not be far better to actively cooperate in developing the new geography which today stands as the much needed interpretative link between the social and physical sciences? Instead of regretting a passing order, would it not be far worthier to assume the task of developing the new affiliations of geography?

Anonymous, 1927, pp. 54–55. Fair play; high moral standards and civic and personal responsibility are to be developed. The pupil is to recognize and strive to attain the best in human behavior. He must respect law and reverence God. Neither an individual nor a nation can rise higher than its ideals. Geography tells of the home life; the community; the school and church and how each have helped in the advancement of society. The work of the explorer and missionary is praised. Industrial activities embracing labor, thrift and cooperation are seen to be the foundation of national greatness. The nations of greatest power are seen to be those that have revered learning, respected law and worshipped God. The ideal of democracy tho [*sic*] far from perfect, is the world's best effort to promote the well being of all. Geography can prove to the pupil that self-government is founded on morality and education, and that rights involve duties.

Mayme Pratt Renner, 1929, p. 292. A geography textbook is, at best, but a gaunt skeleton whose only purpose is to give form and dimension to the geography course.

John E. Orchard, 1930, p. 187. Economic geography is a newcomer in the college and university. Very few courses date back more than twenty-five years and, according to my information, it was not until 1919 that the first professorship in economic geography was established in an American university. Tho the progress in the recognition of the field has been rapid and substantial, economic geography has not been fully accepted into the university curriculum.

John E. Orchard, ibid. A course in economic geography was introduced recently into one of the more liberal schools of social thought in New York City. It languished and attracted few students. Finally, the name of the course was changed from economic geography to descriptive economics. No change was made in subject matter or in the instructor, and now the course prospers. In fairness to the field, it should be stated that the instructor has had no training in geography.

V. C. Finch, 1930, pp. 181–182. I take it as almost axiomatic that work for a liberal arts degree should not have technical information as its principal content. This rule certainly applies to an introductory course in geography in the liberal arts college. The objective in such a course should not be information alone but also a way of thinking about the world. Such a course should provide not merely information but also a systematic method for the arrangement and correlation of facts of geographic significance. It should open a field for later enrichment by study, if the student goes on to subsequent courses in geography, or by the experiences of life in case he does not. It should afford a foundation upon which he may build a broad tolerance and sympathetic appreciation of the peoples and problems of other lands than our own. It should be designed to generate a type of thinking that is the antithesis of provincial.

Charles S. Preble, 1931, p. 125. After all, the teacher is the vital factor in the use of visual aids.

Alfred W. Abrams, 1931, p. 145. When we refer to visual aids, we are thinking for the most part of pictures, altho other objective aids might be included.

Isabel K. Hart, 1931, p. 172. The parts of the world where life is still primitive offer the best choices for our journey studies, in my opinion.

Katheryne Colvin Thomas, 1931, p. 247. A unit in mathematical geography . . . is presented here . . . [in] an attempt to meet the problem of individual differences among students in a required geography course in a teachers college.

Harlan H. Barrows, 1931, p. 353. The chief end in teaching geography is not information, but ability to think geographically. The outstanding educational objective of geography, in other words, is to help make purposeful thinkers and successful doers, not to create animated gazetteers. In order to think geographically, pupils need something suitable to think about, and inducement to think, and appropriate guidance in thinking. The attempt to meet these requirements gives rise to numerous problems in teaching.

Pauline Rafter Powers, 1932, p. 171. In a fifth grade on the opening day of school last fall, I found, by means of a false and true test, four pupils who did not know on what continent they lived, one unable to tell in what country he lived, three who did not know the name of their state, and two who were ignorant of their city's name. Fourteen pupils could not state the direction in which Canada lies from the United States; six were in error about Mexico's location, eighteen knew nothing of the existence of Central America, and five had never heard of Alaska.

Helen M. Kopf, 1932, p. 194. We have one school of geographers who state we should follow the pedagogic maxim of "from the known to the unknown" and begin with the local or home geography. Another school of thinkers in the geographic field, holding to another pedagogic maxim, from the simple to the complex, believe we should first take up the study of people living simple primitive lives on some part of the globe and later take up the complex life of our own locality and civilization, so that our pupils may make the contrast, and better appreciate the present. They believe present-day life in the home community is far too complicated to be understood by the average young child.

L. Dudley Stamp, 1934, p. 121. There is a wide agreement among educators that at some time during a school course a place should be found for the study of the home area and the home environment.

K. C. McMurry, 1936, p. 98. In order to develop the minimum degree of technical competence which is required for the necessary work, the field training of geographers will need to be expanded to many times its present average offerings in geography departments.

J. R. Whitaker, 1937, p. 50. The common lack, in geography, of a cumulative building up of facts and ideas grows in part out of the very nature of the subject. Instead of each topic or problem resting firmly on the preceding ones, it is more often true that each, tho related to preceding ones, stands more or less by itself. In the words of an observant college student, it is like climbing up and down a row of step ladders.

J. R. Whitaker, ibid., pp. 51–52. Another merit of city study as a culminating topic in economic geography is the ease with which it can be carried out. Materials are commonly available for cities even tho lacking for the region to which they belong—photos, street maps, trade statistics, historical accounts, Chamber of Commerce publications, federal commerce reports, port descriptions. The student who takes a city as a special topic on which to report is certain of some measure of *success* in his search for materials.

Rafael Picó, 1941, p. 292. Davis has been called the father of American geography and in view of his writings, teachings and personal influence on present workers in the science, he still merits that distinction.

Carl O. Sauer, 1941, p. 4. ("Foreword.") The business of becoming a geographer is a job of life-long learning. We can teach a few skills such as the making of maps of various kinds, but, mostly, in the instructional period, the best we can do is to open doors for the student.

J. Russell Smith, 1942, p. 316. Consider these facts and you will do two things: (1) be lenient in your judgments of primitive peoples for their apparent stupidity; (2) put the illustrations in your next book in the most easily findable manner by referring to the page on which the illustration is to be found.

J. R. Whitaker, 1943, p. 21. Lastly, we must keep the faith. Man has been likened to a fish in an aquarium, unable to avoid using up the precious oxygen on which its life depends. The comparison is suggestive but hardly adequate, for man can do much to balance resource depletion by restoration. It is on the knowledge that man can, in large measure, maintain the quantity and quality of his natural resources and on the faith that he will have sufficient good sense and self-control to do so that the conservation teacher works.

Henry J. Warman, 1944, p. 305. We see many new "Air Age" maps, distances measured in hours instead of miles; we see the probable air routes of the future—some crossing, and even terminating in, desert wastes. We must grant that "Global Geography" has great appeal and that the airplane has probably done most to make it so in the eyes of the high school student.

Kirk Bryan, 1944, p. 183. To the beginner and the immature the categories of knowledge are a necessary scaffolding for the building which is eventually to contain his realization of geography as a single entity.

George T. Renner, 1944, p. 321. America, almost alone among the great nations, has been accustomed to disregard geography as a major field of scholarship, and has relegated it to the position of a child's subject in the elementary school rather than one of the foundations of education. This practice has left the American people and their leaders peculiarly inept at understanding events and relationships in terms of the world in which they live. This would be ludicrous if it had not proved to be so nearly disastrous to the nation.

Walter W. Ristow, 1944, p. 331. Something new has been added to geography! This is hardly a novel experience, for a number of *new* geographies have been conceived and born within the life span of many members of the profession. Looking back, we can recall the colorless *location* geography still current around the turn of the century, *physical* geography popular in the first decade of the nineteen hundreds, and *human* geography which emerged in the years preceding the first World War. By 1915 *economic geography* was gaining a position which has been maintained, at some educational levels, virtually down to the present.

J. R. Whitaker, 1944, p. 281. An army aviation student was sitting on the front row of the class, looking at a large globe. Suddenly he leaned forward and said, in all seriousness, "That globe is wrong. There simply isn't that much of the Pacific Ocean." This was an extreme example of the geographical ignorance of our young men.

Eugene Van Cleef, 1944, p. 246. Anyone aspiring to devote his energies to international relations, whether trade, political, religious, journalistic or any of the many other phases [*sic*] can hope to achieve total preparation only if he has had a good geographic education along with the usual techniques applicable to the particular field in which he specializes.

Richard Joel Russell, 1945, p. 301. Geography was one of the earliest interests of literate man. The tradition of Herodotus, the study of peoples and places, has found champions in every age. The scholar of post-Columbian days ordinarily chose to be pictured with a globe at his side. Nor has respect for geography decreased during modern times. Public need for the field has been demonstrated many times by the way that schools have admitted it to their curricula.

Richard Joel Russell, ibid., p. 302. To an older generation geography became boundaries, capitals, rivers, capes, and straits. The idea of rôte was somehow more closely identified with geography than with the kings, presidents, generals, wars, and dates of history. Practically everyone despised the subject.

Richard Joel Russell, ibid., p. 312. The broader program of public enlightenment must proceed from the colleges, thru well-trained and firmly-grounded grade-school teachers, to the public at large. Widespread adoption of "global" as a slogan has served useful purposes in bringing geography to public attention, but only sound educational programs, pursued with vigor and patience for many years, can raise the general level of geographical knowledge above the discouraging plane it occupies today.

J. Russell Smith, 1947, p. 101. The geography book contains a great mass of facts culled by the author from a much greater mass of facts. Does any teacher expect a pupil to learn all the facts in the textbook? Do your worst, but it is not likely that all the facts will be remembered, even for a day. Since few will remember geographic facts very long, why then burden the child at all? Why make a book loaded with facts?

Eugene Van Cleef, 1947, p. 92. About twenty-five years ago when the so-called social studies began to make a favorable impression upon our school authorities, there was a grand rush by some geographers to classify geography as a social science. They feared geography would be thrown out of the curriculum if not so defended. Whether or not these ardent persons were engaged in mere political strategy or were firm believers in what they argued is not certain and may not be important now. The fact remains that in the course of the years since passed, great numbers of geographers have become converts to the idea, and this, I think, has been most unfortunate. That movement to "save" geography as it were, sincere as it may have been was a great mistake notably retarding appropriate recognition of the value of the science by the general public and by our colleagues in the realm of the physical sciences. A world war was necessary to regenerate geography, an expensive procedure, and no one can be certain how long the effects of this stimulus may last.

Richard Hartshorne, 1948, p. 125. It should harm no one to be thrown from his intellectual horse—however high the horse—on the contrary, the experience may be taken as a part of a continuing education.

George Beishlag, 1949, p. 60. When geographic writing becomes easy to read [*sic*] we will find the public will want more of our geographic interpretations.

Lawrence Martin, 1950, p. 175. No college professor ever taught me so much or did so much good as William Morris Davis did. He pulled me up by the roots, pruned me, fertilized me, and set me out again in the garden of geography.

Kirk Bryan, 1950, p. 196. To William Morris Davis the modern American geographer owes tribute. He preached the gospel of geography on all occasions during his long and active life: geography in the elementary schools, in the high schools, in the colleges, and in the universities. As an advocate he had no peer. Devoted and dedicated to the analysis of land forms, he also spent much time and thought on the methods of teaching the elements of physical geography: the earth in space, its diurnal and annual rhythms, the tides, weather phenomena, and other elementary facts and processes of our home, the earth. Moreover, he continuously advocated investigation of the relations of all the primary factors of physical geography with the life of man. He was consistently friendly to the cause of "human geography."

George H. T. Kimble, 1951, p. 46. The teaching of geography promotes an awareness of the true nature of man's environment. For want of this far too many of our fellows are still treating the world as if it were a bank to be burgled. Until we come to treat it rather as a trust to be husbanded, we shall make little progress with the task of bringing the habitable part of it into a state of enduring fruitfulness.

Lois Wheeler, 1952, p. 281. The learning and teaching of third grade geography can become an enchanting experience for a class of children if they are lucky enough to have an enthusiastic teacher who has access to some teaching materials to illustrate the subject.

J. Russell Whitaker, 1954, p. 231. I am sure we all feel the satisfaction that comes from seeing a task well done. We can say to ourselves, "this much has been accomplished. Here are landmarks to indicate the distance that we have travelled."

J. Russell Whitaker, ibid., p. 232. I would like to sound a warning prompted by a weakness long evident in the thinking of American geographers. Having arrived at a broad, challenging view of the scope of our field, we should resist most sturdily any effort to shrink it and impoverish it. Those of you who are familiar with the history of the efforts in North America to delimit our field realize that too commonly individual geographers have tried to force this broad field into the narrow frame set by their special aptitudes and interests. It has been difficult for some to define the field save as outlined in their own mental image. If leaders have avoided this pitfall, some of their followers have not succeeded in doing so, hence the succession of dogmas and shibboleths that has plagued us.

J. Russell Whitaker, ibid., p. 238. I dare say that we geographers have never learned to make truly effective use of our great men. We have ignored them perhaps. . . . If we have not ignored them, we have worshiped them uncritically, or have allowed them to hamper and to bind us. Great men can actually cost more than they are worth if we follow as blind copyists or if we allow the examples they have set to dampen our ardor for further investigation and writing or to make us belittle our own efforts.

J. Russell Whitaker, ibid., p. 242. In this big camp of diverse tasks there is room for a wide variety of interests and capacities: for the academic mind, and the practical; for the person who must visualize, and for the person who likes to deal in abstractions; for the historical mind, and the mathematical; for one who must analyze a situation down to its minutest elements and for the person who is gifted at seeing wholes; for the writer adept at description, and for the man who loves exposition but who has not the gift of visualization. There is room here for geographers who delight in taking pictures, and for those who prefer to draw maps; for the worker who is unhappy unless he is outdoors, and his fellow worker who is never so near heaven as when he is buried in a document room; for the person who is satisfied only with the actual concrete details of earth location, and for the one who quickly forgets details, but is constantly searching for generalizations; for the geographer who prefers to study in the home of his boyhood and youth, and the one who is most happy when studying in foreign lands; for the person who is most attracted by unknowns in contemporary geography, and for the student who is happy to reconstruct the past. There is a place, too, for the man who is limited to the use of his native language, and for the scholar who is gifted at serving as an intermediary between those of his own language and those whose thoughts require translation. We have need for the geographer of expansive range and prodigious memory, who writes voluminous definitive works, as well as for the one of limited energy, but critical bent, who is a sharpshooter, who is on the lookout for repeating associations and sequences, and for the worker who is most impressed with the unique, the non-repetitive, in his study of geography.

J. Russell Smith, 1954, p. 84. The members of the Congress of the United States carry our lives in their hands. Think of the danger we are in by having a government in which a majority of the law makers are geographic illiterates. No person should be allowed to enter the primaries for election to the American Congress who had not already passed the geographic examinations necessary for admission to the United States Consular Service. Certainly the law maker needs to know as much as the cub consul.

Carl O. Sauer, 1956, pp. 296–297. The training of the geographer should give attention finally to the history of geographic thought, to the ideas that have prompted and focussed geographic inquiry, and to the circumambient intellectual climates within which geography had lived at different times and places.

Carl O. Sauer, ibid., p. 299. There is nothing wrong with academic geography that a strong coming generation cannot take care of. We can have the needed succession if we free it as much as we may to do what each can do best and wants most to do. It is not for us to prescribe by definition what they shall work at or by what method they shall do so. Academic freedom must always be won anew.

Gilbert F. White, 1958, p. 6. At least three problems arise in arranging programs of study for entering graduate students in geography. One is that students come with widely different formal preparation in the elementary disciplines of geography and related fields. A second is the wide spread in command of skills among those who have had the same formal courses. A third is the uncertainty of new students as to the areas in which they wish to specialize.

John R. Borchert, 1960, p. 15. An educational blind spot, such as the one we have had with respect to geography in the United States, tends to be self-perpetuating; but it cannot be entirely so. Innovation comes slowly at first; then, if it is sound, it gathers speed even though the innovators tend partly to negate their efforts through inexperience and uncertainty in the early stages. Such has been the story with modern geography in the United States. Meanwhile, there is no panacea. There is no easy substitute for thousands of man-years of effective teaching, research, articles, and books—both popular and professional. The opportunity for, and the ability of, geographers to reach large numbers of people in schools, the general public, and management is greater than it ever has been before, and it continues to grow. That opportunity and ability depend upon the product, not the label.

David Lowenthal, 1961, p. 245. Unable to organize objects in space, to envisage places out of sight, or to generalize from perceptual experience, young children are especially poor geographers.

William D. Pattison, 1962, p. 367. This promises to be a banner year for geography in the American high school. With the opening of the 1962–1963 academic year, the High School Geography Project inaugurated a period of programmed experimentation in selected classrooms across the country.

Andrew H. Clark, 1962, p. 232. The road of scholarly endeavor and professional achievement has no room for the mediocre, the lazy, the indifferent, the complacent, or the pretentious. Progress along the way asks everything you have to give, but it is not without its inns of refreshment and chapels of ease.

George Kish, 1963, p. 602. In the past few years, criticism of American geography has taken on a menacing aspect, resulting in the demise of geography as an academic discipline in several major institutions of higher learning.

John K. Wright, 1963, p. 4. I could talk all day about wild geographers I have known—their haunts and habits and their various kinds and varying degrees of "wildness." May we always have wild geographers! may they escape domestication or being made spectacles or making spectacles of themselves in zoos or circuses! may they continue to roam at large, galloping freely or browsing happily over the vast wildernesses of geographical ignorance!

George Kish, 1964, p. 1. American geography is facing a decade of unprecedented change in an academic environment of expanding horizons. The constantly growing demand for qualified teaching staff and the increasing enrollment will place additional burdens on members of the profession both at the graduate and undergraduate levels. It is against this background, one of an expanding educational environment, that any forecast of new directions in American geography must be made. We have to prepare our projections, partially on the basis of past performance, partially on the basis of what we may feel will be necessary for American geographers to do in order to enhance the standing of our profession in the academic community.

George Kish, ibid., p. 2. The coming decade in American geography will be, if past experience is any guide, more demanding than that immediately behind us. Enrollment increases in the United States establishments of higher learning are inevitable. There will be more students in more classes, demanding more textbooks and calling for the services of more teachers. Pressure on every geographer will be double: pressure at the undergraduate level, for more and larger classes, and for the writing of more textbooks; pressure at the graduate level for more time to be spent in graduate instruction, counseling, guidance, to satisfy the spiraling demand for qualified teachers of geography. As new problems face the administrators, more advice will be needed, more committees will be created, and more time will be spent on more reports. A proliferation of present tendencies, an aggravation of present shortcomings—this is likely to be our lot.

George Kish, ibid., p. 4. A small but active group of American geographers have pursued, over the years, a somewhat esoteric, yet delightfully rewarding branch of our science, the history of its origins and development.

F. Kenneth Hare, 1964, p. 459. Years ago I resolved two things: that I would never write a methodological address, on the grounds that such addresses make young workers overwatchful; and secondly, that I would plan my work so as to avoid being quoted in other peoples' methodological addresses.

Fred Kniffen, 1965, p. 577. What I should like to see is a dedicated group of young workers who will with all deliberate haste survey the surviving evidence of the oldest occupance forms and patterns, who will supply us with concepts, terms, and usable quantities, who will be inspired, to paraphrase slightly some now-forgotten source, to think and write about these important things, to be less concerned with sharpening methodological tools, with less playing of closed games with themselves, thereby to raise this very fundamental and satisfying segment of geography to the respected place among American fields of scholarly enquiry that it inherently deserves.

Preston E. James, 1967, p. 3. Three words that geographers frequently use, sometimes inexactly, are space, area, and region. Space is boundless and extends in all directions. When geographers use this word they usually refer to earth-space, which is the three dimensional zone forming the surface of the earth including its atmosphere. Area, on the other hand, is a segment of earth-space, and is bounded. As abstracted on a map, area is a two-dimensional geometric surface, but it is usually conceived by geographers as three-dimensional. A region, as defined by geographers, is an area identified by specified criteria.

Meredith F. Burrill, 1968, p. 4. Starting at an early age, people get their connotations of geographic terms by observing entities in real life or in pictures that are referred to by those words by people considered authoritative. The authorities may be parents, other older people, contemporaries who seem to know, teachers, or authors. Observing is followed or accompanied by individual visualization of an archetype. Once a connotation based on such an archetype is adopted it becomes a primary meaning and mental set prevents it from being readily given up.

Raymond E. Crist, 1969, p. 306. In the evolution of living organisms as well as of institutions, freakish adaptations and abortive mutations have occurred. The profession of geography will not be well served by the growth of excrescences, unseemly deformations, and overdeveloped, overspecialized members. Although history does not tell us how to adapt to the future, it does provide us with a necropolis of mistakes which it is not necessary to make again.

William T. Pecora, 1969, p. 73. For a long time I have had some thoughts to express to geographers, and no better opportunity is open to me than here at your national meeting. From my early student days a peripheral contact with your profession has created a profound respect for its intellectual quality, and I could cite a long list of your colleagues who have influenced my thinking over the years.

J. M. Blaut and David Stea, 1971, p. 393. Geographic learning is clearly if not completely distinguishable from social learning. Here is a very good argument for the distinctive function of geography as a school subject and science. This special function may emerge from some ancient and primitive "sense of place," or some abyssal "protogeography." But it is discovered anew whenever a child plays hide-and-seek, or loses his way, or stubs his toe.

Norton Ginsburg, 1972, p. 5. Responsiveness and responsibility also assume an understanding of what to respond to from *outside* the scholarly society. Such a society, in the nature of the case, reflects the standards and values of the broader society of which it is a part. On the other hand, as a community through which collective wisdom can be pooled, expanded, and applied, it also may serve to clarify those values and advance those priorities which may be at variance with general custom and consensus. Thus, we arrive at the question of "relevance," that totem word which provokes such mixed reactions.

William W. Bunge, 1973, p. 331. ("The Geography.") The Geography is a unit. The Geography contains a certain set of concepts; a finite, a learnable collection. The conceptual content of geography, as opposed to its factual content, can be learned in perhaps three years. Yet geography is such an exceptional science that it is never taught as a total unit. No introductory course has ever attempted to ground its young students in The Geography, the total conceptual content of the trade. Textbooks are nonexistent in the subject. Even readers, even lists of "The Hundred Great Geographical Concepts," are nonexistent.

Yi-Fu Tuan, 1975, p. 205. In the last fifteen years geographers have shown increasing interest in mental phenomena.

Janice Monk and Susan Hanson, 1982, p. 15. Many geographic research questions apply to both men and women, but are analyzed in terms of male experiences only.

Larry R. Ford, 1982, p. 131. Once more, geographers have begun to worry about the future of the discipline. Rumors of departmental cutbacks and eliminations abound, and a softening of the job market for graduates is predicted. Once more we are wondering what can be done to revitalize the field and finally get some respect.

Michael Libbee and Thomas J. Wilbanks, 1982, p. 382. A department can urge its university to conduct solid, consistent evaluations of all programs and take necessary actions so that, when it is evaluated, it will rank with the best in the university.

Nicholas Helburn, 1982, p. 445. Intrinsic and crucial to the quality of my life is the sense that by my living I leave the world a better place than I found it. I suspect many of you have a similar ideal. (In times of discouragement, like these, one may have to rephrase it to say that the world would have been even worse but for my contribution.) Each of us makes a modest contribution, but the faith is that together we make a positive difference. Insofar as geographers contribute as policy makers and advisors, as citizens, as managers, and as consciousness raisers, the "quality of life" policy goal is especially pertinent to the personal ideal of leaving the world a better place.

Nicholas Helburn, ibid., p. 447. In our teaching, both formal and informal, we spread the knowledge of process and its connection to place, we share an ecological way of thinking, heightening the relational view of events and things; we contribute a model of the world into which our students can fit data and generalizations. From all of this they derive deeper understanding and thereby lead richer lives.

A. David Hill, 1982, p. 1. Good teachers and good problem solvers, and the two most basic problems they face are what and how to teach—what might be called the "content" and "process" problems. Good teachers constantly address and make new responses to both problems; they are an unsatisfied if not frenetic lot. Although teachers may see the content and process problems as inextricable, most published discussion about education by geographers focuses only on the content problem.

A. David Hill, ibid., p. 2. Efforts to teach undergraduates marketable skills and to provide internships and counseling so that they might be more vocationally competitive should be encouraged, but they must not overshadow efforts to improve the effectiveness of our role in liberal education.

A. David Hill, ibid., p. 3. If learning the subject matter of disciplines is the single goal of liberal education, then concerns about various teaching methods (e.g., lecture, discussion, simulations, self-pace tutorials, programmed instruction, etc.) are relatively unimportant because numerous studies have shown no difference among distinctive methods of instruction in a given subject when evaluated by student performance on examinations of subject-matter knowledge. In this sense, our debates over "traditional" versus "nontraditional" teaching methods seem basically insignificant.

A. David Hill, ibid., p. 4. Our goal as liberal educators is to provide learning conditions and experiences that help students develop liberating qualities—certain cognitive styles, skills, and socio-psychological attributes—as well as learn geography.

Avinoam Meir, 1982, p. 6. Results of a recent survey reveal that a course on the history and philosophy of geography at the undergraduate level is not offered nor desired by the majority of American and Canadian geography departments or faculty members. In the view of the present author, this situation threatens the capability of the discipline to educate and prepare properly the future generation of geographers, thus jeopardizing the very existence of the discipline.

Avinoam Meir, ibid., p. 7. We must regard the course on history and philosophy of geography as the gate through which we send all of our students to the outside. Having established a self-identity with their discipline, students will be capable of becoming ambassadors of good will.

Avinoam Meir, ibid., p. 7. To establish a professional self-identity within our students we must view the professional personality as being similar to the human personality. As members of social, cultural, or national groups, human beings are taught their national or ethnic history in order to increase their awareness of the group and thus to reduce alienation within the group and within society as a whole. Such a process should be applied in geography as well, to enable students to relate as members of the geographic community and to geography as part of the larger scientific community.

Daniel W. Gade, 1983, p. 261. It seems clear that the knowledge and use of foreign languages are now peripheral to the work of most American geographers.

Daniel W. Gade, ibid., p. 263. The American textbook assumes that undergraduate students are incapable of reading anything in a foreign tongue. They also convey the false message that geography owes nothing to scholarship outside the English-speaking world.

Daniel W. Gade, ibid., p. 264. The richest geographical literatures in the world are essentially ignored on this side of the Atlantic. German, in which is recorded the lion's share of our developmental heritage as a discipline, is not readily accessible to most American geographers in the late twentieth century. French is somewhat better known, yet francophone geographers are rarely cited by Americans for their outstanding contributions. The outpouring of materials in Russian is wasted on most of us to a considerably greater extent than those in French or German. Could it be that the heed paid to books and articles from the United Kingdom reflects linguistic accessibility rather than any superiority of British geography over that from Germany, France or the Soviet Union? Except for a handful of regional specialists, the geographical literatures in Japanese, Arabic, Italian, Portuguese, or Dutch are not even in the realm of potential retrieval for most American geographers. Only Spanish has much visibility among us as a field-oriented research language, thanks to a recently flourishing, though now declining, Latin Americanist specialization.

Richard Morrill, 1983, p. 3. Personally, I am very concerned with what I view as a retreat from rigor in the training of many students, including the abandonment of foreign languages, cartographic skill requirements, and statistical methodology.

Bruce Mitchell and Dianne Draper, 1983, p. 9. . . . the discipline as a whole has not shown the type or degree of concern exhibited by cognate disciplines which have extensively debated the appropriateness of ethical codes and certification to improve the handling of ethical dilemmas.

Gilbert M. Grosvenor, 1984, p. 418. The races, tribes, peoples of South Africa, Southeast Asia, India, Southwest Asia, the nature of the places they inhabit, the economies by which they exist, their traditions and values—if we are to act in the world with authority, we must know the world.

## HISTORY OF GEOGRAPHIC THOUGHT

Jedidiah Morse, 1825, p. 1. The present state of the world and the prospects opening before us render the knowledge of Geography a necessary part of good education.

Franz Boas, 1887, p. 138. Geography is part of cosmography, and has its source in the affective impulse. It depends upon the desire to understand the phenomena and history of a country or of the whole earth, the home of mankind. It depends upon the inclination of the scientist towards physical or cosmographical method, whether he studies the history of the whole earth, or whether he prefers to learn that of a single country.

Charles R. Dryer, 1913, pp. 145–151. The history of the development of secondary school geography in the United States during the past century may be divided into . . . six stages, each characterized by some dominant idea: the gazetteer stage, the wonder book stage, the natural theology stage, the general physiographic stage, the specialized physiographic stage, and the biogeographic stage (ontography).

L. A. Bauer, 1914, pp. 481–499. The famous mathematical physicist Stokes is said to have remarked that we must not forget that the chief instrument of investigation, the mind, is itself the object of research.

G. B. Roorbach, 1914, p. 801. There prevails a general impression even among geographers themselves, that there is little or no agreement as to what geography is or what its purposes and problems are. It has been commonly said that there are as many definitions of geography as there are geographers; that the subject is not a distinct and separate science with a single aim and purpose.

Nevin M. Fenneman, 1919, p. 3. It is a peculiarity of geography to be always discussing and debating its own content—as though a society were to be organized for the sole purpose of finding out what the organization was for. This is not said by way of criticism. . . . The situation is, however, unique and can scarcely fail to be remarked by on-lookers from other sciences, who have no such doubts as to what their subjects are about. The basis of this constant concern is not greed but *fear*. Geography wages no aggressive wars and seems to covet no new territory. In certain quarters it bristles with defense; but it is mainly concerned with purging its own house rather than spreading its borders. To rule out "what is not geography" would seem from the discussions to be much more important than to find and claim geography where it has been passing under other names. The constant apprehension is that by admitting alien subjects we will sooner or later be absorbed by a foreign power and lose our identity.

Charles Redway Dryer, 1920, p. 14. Geography is one of the oldest of the sciences, but it has had a long adolescence and is still among the youngest in some aspects of development. It is a white-haired centenarian with the speech of a child of ten. It did not and could not develop beyond the stage attained by the physical and natural sciences. Although the mother of half the sciences in existence, geography is dependent upon her daughters for support.

Carl O. Sauer, 1921, p. 130. ("Regional Economics.") There have been numerous discussions of the scope of geography, and especially there have been examinations of the periphery of the science. Much less attention has been given to the determination of particular objectives within the field of geography. Geography is suffering from a scattering of interests over too broad a field for the limited number of workers engaged in it.

Harry Elmer Barnes, 1921, p. 330. In no country has the relation between geography and history been more thoroly accepted than in the United States, partly due to the connection between American and German anthropogeography, and partly to the more evident connection between political and geographical history due to the process of westward expansion.

Roderick Peattie, 1923, p. 279. If one understands how a people came to a state of culture and why they think, act, and have their being as they do, one comes almost invariably to sympathize with their condition. Geography is in this way a sympathetic sociology.

G. T. Renner, Jr., 1926, pp. 267–268. ("Geography's Affiliations.") In the early Greek, Moorish, Italian and Portuguese schools, geography was at different times astronomical, mathematical, descriptive or cartographical in nature. To the Polos, to Columbus, to the Venerable Bede, to Stanley or to Stefansson, geography has meant travel lore, and this explorational aspect of geography continues down to the present day.

In France, geography was arrived at thru the study of history; Ratzel, in Germany used it as an adjunct to anthropology. In the United States, Dutton, Gilber, Powell and others developed under the name of geography, a new science of genetic or evolutionary physiography. This was greatly expanded by Davis and his contemporaries and has been carried to a logical perfection and completeness by Professor Douglas Johnson in such works as "The New England-Acadian Shoreline."

Alexander McAdie, 1929, p. 38. We define geography as "the curiosity of mankind concerning its whereabouts."

Albert Perry Brigham, 1929, p. 62. This Association grew out of physiography, or, shall we say, physiographers? Then we swung to human geography, and perhaps became too free with the phrase, geographic influence. Now we are coming back to symmetrical views and modest caution. The face of the earth and the distribution and adjustment of life are all geography. This is the most inappropriate time in all history to be concerned about the interfering or overlapping of any, or all of the sciences. Notwithstanding our prejudices and programs, geography in research and geography in education will take on what it wants, and slough off what it does not want.

G. T. Renner, 1930, p. 344. Geography, the "Mother of Science," has given birth to half a score of daughter sciences which possess clear-cut fields of subject matter, but its own field has been illy defined. This is perhaps the result of the delightfully vague name, "geographia," which Eratosthenes of Alexandria (276–194 B.C.) bestowed upon it.

W. S. Dakin, 1931, p. 294. The original source of geographic information is travel.

Albert Perry Brigham, 1932, p. 49. In the Nineteenth Century, studies in mature geography were mainly incidental to inquiries in other fields.

W. M. Davis, 1932, p. 213. Let no geographer therefore feel himself peculiarly conditioned because his subject is composite. For various other subjects are also composite. Few indeed are not.

Wallace R. McConnell, 1933, p. 151. Geography is not a study of the earth as the home of man; it is what one thinks about as one studies how human life adjusts itself to the natural world. Geography is not a study of relationships; it is what one thinks about as one lists or catalogs relationships. Geography is not what we read or what we see or what we hear; it is only what we think.

William J. Berry, 1933, pp. 236–242. There are reasons why geography may properly adhere to both the social and natural science groups. Certainly most of the work now done in geography is closely akin to the work of the social scientists. It would seem equally certain that the geographer has more to offer to the student of the social sciences than to the student of natural science, yet descriptive climatology, descriptive geomorphology, and other aspects of the field of natural earth that come within our proper sphere are essentially natural science. If we may officially join the social scientists without being read out of the natural science group, I think the change would be most advantageous. It is my belief that the dual relationship of our field is in theory defensible.

Clifford M. Zierer, 1933, p. 239 (quoted in William J. Berry, 1933). Geography is concerned primarily with human activities and their relationship to natural environmental conditions. That viewpoint enriches the study of social sciences very greatly and the contributions of geography are fully appreciated by many workers engaged in history, economics, and political science.

It is true that geography deals with facts belonging to the realm of the natural sciences. However, it is not the goal of most geographers merely to discover and explain bare physical facts. It is more often true that the geographer accepts the facts given by specialized workers in the natural sciences and uses them in this interpretation of human affairs. Geography, therefore, contributes less new material to the natural sciences.

Ellsworth Huntington, 1933, p. 241 (quoted in William J. Berry, 1933). Geography by its very nature combined two things. It is preeminently a science which attempts to discover the relation between the facts of nature and the facts of human society. Therefore to call it either a natural or a social science is a misnomer. Nevertheless, as a practical matter of policy, it is often necessary to classify it in one way or another. Twenty years ago under the influence of Professor W. M. Davis the pendulum had swung much too far toward the geological side and many people thought that geography was primarily one of the natural sciences. Now, however, we are in danger of forgetting that there can be no sound geography without a deepseated knowledge of natural science. Nevertheless, *human* geography in its various phases, such as economic, political, and social geography, is unquestionably the most interesting and practical part of the science for the great majority of people. The workers in human geography, as a rule, find greater sympathy among sociologists, anthropologists, historians and the like than among geologists, botanists and biologists. Therefore, if the subject must be placed in one group or another in any given university, it seems to me in general that it should go with the social sciences.

Nevin M. Fenneman, 1933, p. 241 (quoted in William J. Berry, 1933). In a very general way, geography is dependent on the natural sciences and itself underlies the social sciences. It will stagnate without the former just as the latter will stagnate without geography. The association of geography with geology in the United States has, on the whole, been fortunate. Having its roots in the ground does not prevent flowers and fruit at the top.

R. H. Whitbeck, 1933, p. 241 (quoted in William J. Berry, 1933). I would say that the vary nature of the subject matter of geography places geography in both the category of natural sciences and of social sciences. It is manifestly impossible to change the nature of the subject matter of geography which is established by a century or two of tradition in America. Physical geography is a natural science; economic geography is a social science and geography is both.

Derwent Whittlesey, 1933, p. 237 (quoted in William J. Berry, 1933). I suppose all workers in geography recognize that human geography involves much more observation than the other social sciences and somewhat more opinion than the other natural sciences. Nevertheless if I press the reasoning to its conclusion I always decide that its value as an observational discipline far exceeds its value as a compilation of opinions.

William H. Haas, 1934, p. 53. Knowledge comes but wisdom lingers. The accumulation of information in any science is a relatively simple process and has many followers; but the discernment of the true and the right in their relationships is far more difficult and is reserved to a much smaller group. Specialization to some means an ever concentrating and narrowing down of interests to some single phase of a subject; to others it means ever broadening fields in research interests with a limited goal ahead. Among opinions solicited . . . from some of our leading geographers there is a fairly wide range of opinion relative to foundations and limits in geography in their relation to sound geographic thinking.

A. E. Parkins, 1934, p. 225. The basic objective of geography is a description of the face of the earth. This is what geography means to the layman and what geography has meant since the earliest time; and concerning this general definition there should be no argument. Description of the face of the earth involves many things. Obviously, it is necessary in some way to choose a limited objective within the general field. It is the way in which a geographer chooses to limit that objective that determines what subdivision he pursues. People who choose to contribute to the general knowledge concerning the character and face of the earth by a discussion or study of one of its features are none the less geographers even if we must describe these individuals with such names as geomorphologists, plant geographers, economic geographers, etc.

A. E. Parkins, ibid., p. 230. May we come to think of the function of geography as did Strabo nearly 1900 years ago, who declared that knowledge of this subject "makes him who cultivates it as a man earnest in the great problems of life and happiness."

J. Russell Smith, 1935, p. 20. The subject Geography is now about half way through a half century of handicap because of a simple fact of nomenclature. Almost every worker in geography suffers at some time or other because geography is an old word with a new meaning.

W. L. G. Joerg, 1936, p. 641. Modern geography may be said to date essentially from the establishment of the subject in the universities. In the four countries of Europe that have been the leaders in this development these dates are approximately as follows: Germany, early 1870's; Italy, late 1870's and early 1890's; France, about 1890; Great Britain, late 1890's.

W. Elmer Ekblaw, 1937, p. 214. ("Attributes.") All place relationships or attributes of land forms, or rainfall, or ocean currents; of maize, or micro-organisms, or soils; of man, or language, or even of farm walls and fences; all or any of these fall within the proper definition and field of geography. We might with equal propriety speak of the geography of man; of the geography of soils; of the geography of disease; of the geography of corn; or of highways and bridges even. The criterion by which we decide whether or not a fact or principle, or any body of knowledge relating to anything, is geographic, is whether or not it involves the concept of place and its attributes. The three essentials in every situation, in every drama, are the time, the place, and the thing; and as time implies history, so place implies geography.

Isaiah Bowman, 1938, p. 10. The edge of the world troubled the precursors of Columbus. As exploration advanced, the edge withdrew: it was only in men's minds! It has been retreating ever since.

L. Carrington Goodrich, 1938, p. 400. The Chinese have had a long and honorable history as geographers and map makers.

V. C. Finch, 1939, p. 1. It is not geography but our thinking about it that seems to be driven into frequent reorientations.

Richard Hartshorne, 1939, p. 200. American geography has therefore been markedly influenced by the work of European geographers, but it is notable that this influence has come through but very few contacts—for the most part through Davis, Semple, and Sauer.

Richard Hartshorne, ibid., p. 209. There can be no question that the foundation of geography as a modern science was primarily the work of German students. In whatever country one starts, the study of the development of geography leads backward to the work of Humboldt and Ritter in the early part of the nineteenth century.

Richard Hartshorne, ibid., p. 211. Although the roots of geography, as a field of study, reach back to Classical Antiquity, its establishment as a modern science was essentially the work of the century from 1750 to 1850. The second half of this period, the time of Humboldt and Ritter, is commonly spoken of as the "classical period" of geography. Undoubtedly the extraordinary accomplishment of each of these men, working at the same time but in very different ways, and the influence of their work on all subsequent geography justifies our regarding them as the first masters of modern geography—in that sense as the "founders."

Richard Hartshorne, ibid., p. 639. Geography and history are alike in that they are integrating sciences concerned with studying the world. There is, therefore, a universal and mutual relation between them, even though their bases of integration are in a sense opposite—geography in terms of earth spaces, history in terms of periods of time.

Carl O. Sauer, 1941, p. 3. ("Foreword.") The American geography of today is essentially a native product; predominantly it is bred in the Middle West, and, in dispensing with serious consideration of cultural or historical processes it reflects strongly its background. In the Middle West, original cultural differences faded rapidly in the foraging of a commercial civilization based on great natural resources. Perhaps nowhere else and at no other time has a great civilization been shaped so rapidly, so simply, and so directly out of the fat of the land and the riches of the subsoil. Apparently here, if anywhere, the formal logic of costs and returns dominated a rationalized and steadily expanding economic world. The growth of American geography came largely at a time when it seemed reasonable to conclude that under any given situation of natural environment there was one best, most economical expression of use, adjustment, or response.

A. K. Lobeck, 1942, p. 132. It may be surprising to discover that the man's name George comes from the same root as geography. The Greek root *ge* means the "earth." When Vergil wrote his book called the *Georgics*, he was writing about farming, for the word George means a worker in the earth. Besides *ge*, it includes the root *ergein* meaning "to work," a root found also in energy, erg, metallurgy, and surgeon.

Robert S. Platt, 1948, p. 355. ("Environmentalism Versus Geography.") Geography is "the study of areas with respect to their differences." In this definition what areas are implied? Unmistakenly those of the earth's surface. What differences are to be included? Not microscopic, universal or absolute differences; but those which distinguish homelands from other lands, those same differences that aroused the interest of early travelers to make the first geographical observations.

George T. Renner, 1950, p. 14. Economic geography, or *geonomics* as it is sometimes called, obtained its start during the latter part of the eighteenth century when Immanuel Kant in Germany pointed out the existence of a field of geography which he termed "Mercantile Geography." At about that same time, Adam Smith in England was publishing his *Wealth of Nations*.

George T. Renner, ibid. Today, in America at least, [economic geography] is the dominant aspect of geography.

Edward Coleson, 1952, p. 147. A Gallup survey conducted a few years ago found that two-thirds of the people interviewed could not find Greece on an outline map, nearly half failed to find Spain, and one-third couldn't locate France.

O. Orland Maxfield, 1953, p. 26. Economic geography is that branch of geographic endeavor which probes the distribution of, and relations between, the earth resources and the occupations by which man maintains himself.

Thomas F. Barton, 1954, p. 14. Recently I was asked point blank, "Do you teach that there is an Antarctic Ocean?" My short answer was "No!" In a lecture on "Pattern of the Continents and Oceans" before five hundred students enrolled in a Physical Geography course each year, I stress that scientific research especially during the past few decades has discovered that the term "Antarctic Ocean" is obsolete and therefore we should teach only four oceans.

Lorren G. Kennamer, Jr., 1955, p. 32. [School geography] has changed from a descriptive study to a physical science to a social science in less than one hundred years in this country.

Edward A. Ackerman, 1957, p. 109. The techniques of geography make it something more than a social science.

Clyde F. Kohn, 1959, pp. 122-123. More attention is commonly paid by geographers to the spatial dimensions of economic and political activities than of social or spiritual affairs of man.

Kenneth Thompson, 1960, p. 4. Having long been unhappy with both the connotation and etymology of the term geography as applied to the academic discipline, I find myself in thorough agreement . . . that the field be renamed.

Kenneth Thompson, ibid., p. 5. Criticism of the term geography can be directed along several lines. One objection concerns its derivation from the Greek *gaia*, *ge* (earth) + *graphe* (description). Geography thus finds itself in the anomalous position as one of the few organized fields of study that declares itself nominally to be descriptive. Where is the *logos* in geographical study?

Julian V. Minghi, 1963, p. 4. As the most passionate branch of geography, political geography has not always been free from subjective writing.

Richard J. Chorley, 1964, p. 127. One of the most striking characteristics of geographical analysis which this subject has in common with the other natural and social sciences is the high degree of ambiguity presented by its subject matter and the attendantly large "elbow room" which the researcher has for the manner in which this material may be organized and interpreted. This characteristic is a necessary result of the relatively small amount of available information which has been extracted in a very partial manner from a large and multivariate reality, and leads not only to radically conflicting "explanations" of geographical phenomena but to differing opinions regarding the significant aspects of geographical reality which are worth exploring. Even within a circumscribed body of information there is no universally appropriate manner of treatment, and such treatment is often conditioned either by the general systematic framework which one . . . adopts as an appropriate setting for the information or by the type of question which one is prepared to ask about the "real world." The change in character between the geographical methodologies of the 19th and 20th centuries . . . lies very largely in the abandonment of attempts at causal explanation in favor of functional studies.

David H. Miller, 1965, p. 1. Whether the surface of the earth is regarded from the cultural, physical, or economic side, it is indivisible; geography has an integrated focus.

Terry G. Jordan, 1966, p. 26. Settlement geography represents one special field within our discipline in which American geographers consistently have lagged far behind their European counterparts.

Raymond E. Crist, 1969, p. 305. (Professor Gould was twelve years old.) A study of the history of one's subject tends to make a person not only wiser but more courteous.

Raymond E. Crist, ibid., p. 306. The profession of geography is, or should be, most keenly aware of that hard fact of evolution, *viz*, that everything—people, professions, institutions—must adapt to new conditions or perish. If the geographer fails to think clearly enough about his profession to be able to explain and justify, in short, to communicate, to an educated layman his raison d'être, geography—new or old—will be superseded; for ours is a highly competitive society. However, in the evolution of living organisms as well as of institutions, freakish adaptations and abortive mutations have occurred. The profession of geography will not be well served by the growth of excrescences, unseemly deformations and overdeveloped, overspecialized members. Although history does not tell us how to adapt to the future, it does provide us with a necropolis of mistakes which it is not necessary to make again.

Kirk H. Stone, 1970, p. 5. THE INHERITANCE. In geography's early stages the object of attention was mostly the whole known world. Our professional forefathers struggled for small-scale concepts: how big was the world, was it flat or round, did it curve away from the sun so much northward that it was always cold? It is stimulating to look back on what they tried to do with the limited knowledge they had, but they did it. In time the discipline grew in numbers of followers and took on added significance largely, or at least in major part, because geographers mapped, wrote, and spoke about the whole known world. Geographical goals were to map elements on world-wide bases.

Leszek Kosiński, 1971, p. 615. The geographers of East-Central Europe have had traditional ties with French geography, which persist, and also with German geography during the time when it flourished. The impact of Soviet geography was particularly strong in the early postwar years, especially in Bulgaria, Romania, and East Germany. Contacts with British and American geographers have been developing only during the last decade or so.

William W. Bunge, 1973, p. 331. ("The Geography.") Instead of an accumulation of concepts, what we witness and endure in geography are purges. The Geography of the Sixties, decimates The Geography of the Fifties only in turn to be eradicated by The Geography of the Seventies. Environmentalism was decimated by microfield work and regionalism only to be buried by quantification which in turn is threatened by the geography of the human condition. This is exceptionalism among the sciences. Other sciences accumulate, build, add. The newer physics swallows the older physics, not refutes it. It is time geography does likewise and this can be seen especially with the wave of humanistic geography rising in North America.

Marvin W. Mikesell, 1974, p. 3. The retreat of geographers from environmentalism began in the 1920's and by the end of the 1930's had resulted in nearly total abandonment of the research program initiated prior to the First World War.

Marvin W. Mikesell, ibid. Environmentalism was held by its critics to be merely an hypothesis, subject to verification or rejection, and hence could not be regarded as a definition of the content or objective of geography as an academic field.

Mildred Berman, 1974, p. 11. Had Semple been a man, she would not have had to listen to Ratzel's lectures in an adjoining room with the door ajar when she first went to Leipzig, where women were not permitted to matriculate. Had she been a man, a salary equal to that of her less-celebrated colleagues might have somehow eased the financial strain and discomfort of her last months on earth.

Leonard Guelke, 1974, p. 202. The idea that human geographers ought to attempt to emulate physical scientists in search of theory overlooks the fact that man himself is a theoretical animal whose actions are based on the theoretical understanding of his situation. As man's theoretical ideas change, so will his behavior. Any attempt to describe human behavior in theoretical terms seems doomed. The idealist philosophy gives human geographers a philosophy of explanation that allows them to take full account of the special nature of human theoretical behavior. Rethinking the thoughts of people whose action he wishes to explain enables the investigator to explain human actions in a critical analytical way without theory. This approach is no more subjective than that employed by the positivists. Verification procedures open to critical examination are available to test the worth of any idealist interpretation. The idealist human geographer aims at providing a true explanation of the situations he investigates.

Richard Peet, 1975, p. 564. Marxists theorize that inequality and poverty are functional components of the capitalist mode of production: capitalism necessarily produces inegalitarian social structures. Inequality is transferred from one generation to another through the environment of services and opportunities which surrounds each individual. The social geography of the city is made up of a hierarchy of community environments reproducing hierarchical class structure. Change in the system results from change in the demand for labor. Continuing poverty in American cities results from a continued system need to produce and reproduce an industrial reserve army. Inequality and poverty cannot be eradicated without fundamental changes in the mode of production.

Yi-Fu Tuan, 1976, p. 266. The focus of humanistic geography is on people and their condition. Humanistic geography is thus not primarily an earth science, yet it is a branch of geography because it reflects upon kinds of evidence that interest other branches of the discipline. The following topics are briefly noted from the humanistic perspective: geographical knowledge, territory and place, crowding and privacy, livelihood and economics, and religion. The basic approach to these topics is by way of human experience, awareness, and knowledge. Humanistic geography contributes to science by drawing attention to facts hitherto beyond the scientific purview. It differs from historical geography in emphasizing that people create their own historical myths. A humanist geographer should have training in systematic thought, or philosophy. His work serves society essentially by raising its level of consciousness.

Preston E. James, 1976, p. 2. Opportunity for advanced training in a field called geography in the United States was first offered at Harvard after 1885, when William Morris Davis and his teacher, Nathaniel Southgate Shaler, offered a graduate seminar in physical geography. The paradigm proposed by Davis specified that a scientific study of geography must consist of an "explanatory description" of the earth's physical features and an account of the response of organisms, including man, to these features.

Robert H. Fuson, Harry J. Schaleman, Jr., and Douglas C. Wilms, 1978, p. 319. The first hint of baseball's influence on our discipline came when the chairman of one major league institution (one that offers the Ph.D.) was overheard to tell another, "I'll trade you two cultural geographers with a few more good years left in them for one young cartographer and a draft choice to be named later."

Yi-Fu Tuan, 1978, p. 363. In human geography we sometimes treat people as though they have little or no awareness. This is the approach of sociophysicists. On the other hand, we sometimes postulate a world in which people are always feeling, willing, thinking, and making decisions. This is the natural emphasis of humanist scholars.

Gordon L. Clark and Michael Dear, 1978, p. 356. Over the last decade there has been rapid growth in the number of radical geographers, their degree of organization, and their visibility in the academic media.

Merle C. Prunty, 1979, p. 42. To a considerable degree the successes and failures of American geography since 1950 derive from the nature of graduate training in a surprisingly small number of "leadership" departments.

Edward W. Soja, 1980, p. 207. An increasingly rigidifying orthodoxy has begun to emerge within Marxist spatial analysis that threatens to choke off the development of a critical theory of space in its infancy.

David Sibley, 1981, p. 1. Critical accounts of positivism in human geography, particularly those addressed to the spatial geometry school, have been concerned primarily with its inherent mystifying tendencies, which derive from its theoretical basis in neoclassical economics and functionalist sociology.

Richard A. Walker, 1981, p. 5. As a Marxist, I find myself repeatedly in an uncomfortable position between positivism, which generally occupies a place on the academic right, and the new forms of left "humanist" theory espoused by geographers, such as phenomenology and critical theory. Although I am in sympathy with the latter efforts to create a more methodologically sophisticated and critical geography, I find their shortcomings disturbing. Indeed, much of what passes for "critical" social science seems to be a kind of left-wing libertarianism. This school of thought has as its enemy the vaguest notion of "domination" or social control; as its protagonist, the individual against large collectivities labeled "the state" or "the corporation"; and as its program, strenuous criticism of all positive efforts to comprehend and—dare I say it—manipulate the world around us, but very little else except calls for academic purity and good works.

Barry M. Moriarty, 1981, p. 484. Geographic research tends to adhere either to the phenomenological or positivist approach, the former based upon the study of problems as unique entities and the latter concerned with hypothesis testing and theory development or application. Geographic research is also either basic or applied in its objectives. Basic research in geography is intended to contribute to better understanding of the spatial processes associated with problems and the development of techniques for analyzing and resolving problems exhibiting spatial properties. Applied research uses knowledge about spatial processes and spatial analytical techniques to develop strategies designed to alleviate problems.

Janice Monk and Susan Hanson, 1982, p. 11. Recent challenges to the acceptability of traditional gender roles for men and women have been called the most profound and powerful source of social change in this century, and feminism is the "ism" often held accountable for instigating this societal transformation. One expression of feminism is the conduct of academic research that recognizes and explores the reasons for and implications of the fact that women's lives are qualitatively different from men's lives. Yet the degree to which geography remains untouched by feminism is remarkable, and the dearth of attention to women's issues, explicit or implicit, plagues all branches of human geography.

Janice Monk and Susan Hanson, ibid., p. 12. Marxists have championed social change but, with a few exceptions they have not explored the effects of capitalism on women.

Helen Couclelis and Reginald Golledge, 1983, p. 331. Tasting the fruit of the Tree of Knowledge has traditionally brought sorrows and woes on mankind. As ever more geographers during this past decade were losing their innocence in the arms of philosophy, the age-old curse was upon us once again.

Helen Couclelis and Reginald Golledge, ibid., p. 337. There is, we think, a place in geography for the poet and the prophet and the seer and the critic and the rebel and even the dogmatic doctrinaire, as they all tramp along that intricate network of overgrown and often impassable footpaths that now replace the age-old vision of a royal road to Truth.

Geoffrey J. D. Hewings, 1984, p. 99. Today, it is often difficult to distinguish many contributors in regional science from those in economic geography; the overlap has resulted from the considerable degree of interaction which takes place between civil engineers, regional economists, economic geographers, urban and regional planners, and regional scientists.

# SOURCES CITED


Abrams, Alfred W., 1931. ''Visual Instruction in Teaching Geography,'' *Journal of Geography*, Vol. 30, pp. 145–150.

Ackerman, Edward A., 1937. ''The Wine Valley of North Portugal,'' *Journal of Geography*, Vol. 36, pp. 333–353.

Ackerman, Edward A., 1938. ''Albania—A Balkan Switzerland,'' *Journal of Geography*, Vol. 37, pp. 253–262.

Ackerman, Edward A., 1945. ''Geographic Training, Wartime Research, and Immediate Professional Objectives,'' *Annals of the Association of American Geographers*, Vol. 35, pp. 121–143.

Ackerman, Edward A., 1957. ''Resources for the Future, Inc. and Resource Use Education,'' *Journal of Geography*, Vol. 56, pp. 103–109.

Ackerman, Edward A., 1962. ''Public Policy Issues for the Professional Geographer,'' *Annals of the Association of American Geographers*, Vol. 52, pp. 292–298.

Ackerman, Edward A., 1963. ''Where Is a Research Frontier,'' *Annals of the Association of American Geographers*, Vol. 53, pp. 429–440.

Adams, Cyrus, 1912. ''Maps and Map-Making,'' *Bulletin of the American Geographical Society*, Vol. 44, pp. 194–201.

Ahnert, Frank, 1962. ''Some Reflections on the Place and Nature of Physical Geography in America,'' *Professional Geographer*, Vol. 14, pp. 1–7.

Akatiff, Clark, 1974. ''The March on the Pentagon,'' *Annals of the Association of American Geographers*, Vol. 64, pp. 26–33.

Alexander, Robert, 1964. ''Geographic Data from Space,'' *Professional Geographer*, Vol. 16, pp. 1–5.

Anonymous, 1927. ''Objectives of Elementary Education and How Geography Helps in Attaining Them,'' *Journal of Geography*, Vol. 26, pp. 52–57.

Anonymous, 1951. ''A Plea for Geography, 1813 Style,'' *Annals of the Association of American Geographers*, Vol. 41, pp. 233–236.

Applebaum, William, 1947. ''The Geographer in Business and His Requisite Training,'' *Professional Geographer*, Vol. 5, pp. 1–4.

Applebaum, William, 1955. ''What Are Geographers Doing in Business?'' *Annals of the Association of American Geographers*, Vol. 45, pp. 163–164.

Applebaum, William, 1956. ''What Are Geographers Doing in Business?'' *Professional Geographer*, Vol. 8, pp. 2–5.

Artchinson, Alison E., 1918. ''Physiography as a Basis for Commercial Geography, Botany, and History,'' *Journal of Geography*, Vol. 16, pp. 215–218.

Aschmann, Homer, 1962. "Geography in the Liberal Arts College," *Annals of the Association of American Geographers*, Vol. 52, pp. 284–292.
Atwood, Rollin S., 1957. "Regional Geography in Action, the 'Plan Chillan' of Chile," *Annals of the Association of American Geographers*, Vol. 47, pp. 151–152.
Atwood, Wallace W., 1919. "The Call for Geographers," *Annals of the Association of American Geographers*, Vol. 9, p. 71.
Atwood, Wallace W., 1928. "Research and Educational Work in Geography," *Journal of Geography*, Vol. 27, pp. 263–270.
Atwood, Wallace W., 1935. "The Increasing Significance of Geographic Conditions in the Growth of Nations-States," *Annals of the Association of American Geographers*, Vol. 25, pp. 1–16.
Atwood, Wallace W., 1940. "The Fundamental Basis for the Study of Geography," *Annals of the Association of American Geographers*, Vol. 30, pp. 44–45.
Atwood, Wallace W., 1940. "Geography and the Great Human Dramas," *Journal of Geography*, Vol. 39, pp. 337–343.
Atwood, Wallace W., 1947. "The New Meaning of Geography in World Education," *Journal of Geography*, Vol. 46, pp. 11–15.
Balfour, Andrew, 1923. "Sojourners in the Tropics," *The Lancet*, Vol. 204, pp. 1329–1334.
Barnes, Harry Elmer, 1921. "The Relation of Geography to the Writing and Interpretation of History," *Journal of Geography*, Vol. 20, pp. 321–337.
Barr, William, 1983. "Geographical Aspects of the First International Polar Year, 1882–1883," *Annals of the Association of American Geographers*, Vol. 73, pp. 463–484.
Barrows, Harlan H., 1923. "Geography as Human Ecology," *Annals of the Association of American Geographers*, Vol. 13, pp. 1–14.
Barrows, Harlan H., 1931. "Some Critical Problems in Teaching Elementary Geography," *Journal of Geography*, Vol. 30, pp. 353–364.
Barton, Bonnie, 1978. "The Creation of Centrality," *Annals of the Association of American Geographers*, Vol. 68, pp. 34–44.
Barton, Thomas F., 1939. "The Penguin and the Ostrich," *Journal of Geography*, Vol. 38, pp. 188–191.
Barton, Thomas F., 1941. "Establishing an Inexpensive Weather Station," *Journal of Geography*, Vol. 40, pp. 226–230.
Barton, Thomas F., 1954. "Is There an Antarctic Ocean?" *Journal of Geography*, Vol. 53, pp. 14–17.
Bauer, L. A., 1914. "The General Magnetic Survey of the Earth," *Bulletin of the American Geographical Society*, Vol. 46, pp. 481–499.
Baulig, Henri, 1950. "William Morris Davis: Master of Method," *Annals of the Association of American Geographers*, Vol. 40, pp. 188–195.
Baylor, J. Wright, 1931. "Climate and Man in the Columbia Plateau Area," *Journal of Geography*, Vol. 30, pp. 264–279.
Beishlag, George, 1949. "What's Wrong With [*sic*] Geographic Writing?" *Annals of the Association of American Geographers*, Vol. 39, p. 60.
Bennett, Charles F., Jr., 1957. "A Brief History of Trained African Elephants in the Belgian Congo," *Journal of Geography*, Vol. 56, pp. 168–172.
Bennett, Hugh H., 1926. "Agriculture in Central America," *Annals of the Association of American Geographers*, Vol. 16, pp. 63–84.
Bentley, E. C., Undated. "Biology for Beginners," *Dictionary of Quotations*, London, p. 42.

Bergsmark, Daniel, 1931. "Clermont County—Economic Adjustments to the Environment," *Annals of the Association of American Geographers*, Vol. 21, pp. 111–112.
Berman, Mildred, 1974. "Sex Discrimination and Geography," *Professional Geographer*, Vol. 26, pp. 8–11.
Berry, Brian J. L., 1959. "Statistical Tests of Value in Grouping Geographic Phenomena," *Annals of the Association of American Geographers*, Vol. 49, p. 169.
Berry, Brian J. L., 1964. "Approaches to Regional Analysis: A Synthesis," *Annals of the Association of American Geographers*, Vol. 54, pp. 2–11.
Berry, Brian J. L., 1980. "Creating Future Geographies," *Annals of the Association of American Geographers*, Vol. 70, pp. 449–458.
Berry, Brian J. L., and William L. Garrison, 1958. "Alternate Explanations of Urban-Rank Size Relationships," *Annals of the Association of American Geographers*, Vol. 48, pp. 83–91.
Berry, William J., 1933. "Some Opinions Relative to the Content and Grouping of Geography," *Journal of Geography*, Vol. 32, pp. 236–242.
Bingham, Millicent Todd, 1928. "A Method of Approach to Urban Geography," *Annals of the Association of American Geographers*, Vol. 19, p. 24.
Birdseye, Claude H., 1940. "Stereoscopic Phototopographic Mapping," *Annals of the Association of American Geographers*, Vol. 30, pp. 1–24.
Blaut, J. M., 1962. "Object and Relationship," *Professional Geographer*, Vol. 14, pp. 1–7.
Blaut, J. M., and David Stea, 1971. "Studies of Geographic Learning," *Annals of the Association of American Geographers*, Vol. 61, pp. 387–393.
Boas, Charles W., 1959. "Locational Patterns of American Automobile Assembly Plants, 1895–1958," *Annals of the Association of American Geographers*, Vol 49, pp. 170–171.
Boas, Franz, 1887. "The Study of Geography," *Science*, Vol. 9, pp. 137–141.
Boggs, S. W., 1932. "Boundary Functions and the Principles of Boundary-making," *Annals of the Association of American Geographers*, Vol. 22, pp. 48–49.
Boggs, S. W., 1946. "The Earth and Its Tenants," *Annals of the Association of American Geographers*, Vol. 36, pp. 82–83.
Borchert, John R., 1960. "A Statement Favoring Support of the Term Geography," *Professional Geographer*, Vol. 12, pp. 14–16.
Borchert, John R., 1968. "Remote Sensing and Geographical Science," *Professional Geographer*, Vol. 20, pp. 371–375.
Borchert, John R., 1971. "The Dust Bowl in the 1970s," *Annals of the Association of American Geographers*, Vol. 61, pp. 1–22.
Borchert, John R., 1983. "Instability in American Metropolitan Growth," *Geographical Review*, Vol. 73, pp. 127–149.
Borchert, John R., 1985. "Geography and State-Local Public Policy," *Annals of the Association of American Geographers*, Vol. 75, pp. 1–10.
Bowie, William, 1933. "Status of Geodetic Surveys in the United States," *Annals of the Association of American Geographers*, Vol. 23, p. 42.
Bowman, Isaiah, 1932. "Planning in Pioneer Settlement," *Annals of the Association of American Geographers*, Vol. 22, pp. 93–107.
Bowman, Isaiah, 1934. *Geography in Relation to the Social Sciences*, New York: Charles Scribner's Sons, p. 119.

Bowman, Isaiah, 1934. "William Morris Davis," *Geographical Review*, Vol. 24, pp. 177–181.
Bowman, Isaiah, 1935. "Our Expanding and Contracting 'Desert,' " *Geographical Review*, Vol. 25, pp. 43–61.
Bowman, Isaiah, 1938. "Geography in the Creative Experiment," *Geographical Review*, Vol. 28, pp. 1–19.
Bowman, Isaiah, 1942. "Geography vs. Geopolitics," *Geographical Review*, Vol. 32, pp. 646–658.
Bowman, Isaiah, 1945. "The New Geography," *Journal of Geography*, Vol. 44, pp. 213–216.
Brigham, Albert Perry, 1914. "Early Interpretations of the Physiography of New York," *Bulletin of the American Geographical Society*, Vol. 46, pp. 25–35.
Brigham, Albert Perry, 1915. "Problems of Geographic Influence," *Annals of the Association of American Geographers*, Vol. 5, pp. 3–25.
Brigham, Albert Perry, 1920. "Geography and the War," *Journal of Geography*, Vol. 19, pp. 89–102.
Brigham, Albert Perry, 1922. "A Quarter-Century in Geography," *Journal of Geography*, Vol. 21, pp. 12–17.
Brigham, Albert Perry, 1924. "The Association of American Geographers, 1903–1923," *Annals of the Association of American Geographers*, Vol. 14, pp. 109–116.
Brigham, Albert Perry, 1924. "Remarks on Geography in America," *Journal of Geography*, Vol. 12, pp. 202–205.
Brigham, Albert Perry, 1929. "An Appreciation of William Morris Davis," *Annals of the Association of American Geographers*, Vol. 19, pp. 61–62.
Brigham, Albert Perry, 1932. "Research by American Geographers Since 1900," *Annals of the Association of American Geographers*, Vol. 22, pp. 49–50.
Brooke, M.E., 1929. "Usual Glimpses of St. Thomas, the Island of Blackbeard and Bay Rum," *Bulletin of the Geographical Society of Philadelphia*, Vol. 27, pp. 43–53.
Brooks, Charles F., 1948. "The Climatic Record: Its Content, Limitations, and Geographic Value," *Annals of the Association of American Geographers*, Vol. 38, pp. 153–168.
Brown, Eric, 1975. "The Content and Relationship of Physical Geography," *Geographical Journal*, Vol. 141, pp. 35–40.
Brown, Ralph H., 1930. "The Mountain Communities of the Boulder Region, Colorado," *Journal of Geography*, Vol. 29, pp. 271–287.
Brown, Ralph H., 1951. "The Land and the Sea: Their Larger Traits," *Annals of the Association of American Geographers*, Vol. 41, pp. 199–216.
Brown, Robert M., 1910. "The Change of Response Due to a Changing Market," *Journal of Geography*, Vol 8, pp. 121–128.
Brown, Robert M., 1924. "The Bounds of Racial Geography," *Journal of Geography*, Vol. 23, pp. 41–48.
Bryan, Kirk, 1944. "Physical Geography in the Training of the Geographer," *Annals of the Association of American Geographers*, Vol. 34, pp. 183–189.
Bryan, Kirk, 1950. "The Place of Geomorphology in the Geographic Sciences," *Annals of the Association of American Geographers*, Vol. 40, pp. 196–208.
Bryant, Henry G., 1913. "A Canoe Journey in Southeastern Labrador," *Annals of the Association of American Geographers*, Vol. 3, p. 111.

Bunge, William, 1964. "Geographical Dialectics," *Professional Geographer*, Vol. 16, pp. 28–29.
Bunge, William, 1966. "Locations Are Not Unique," *Annals of the Association of American Geographers*, Vol. 56, pp. 375–376.
Bunge, William W., 1973. "The Geography," *Professional Geographer*, Vol. 25, pp. 331–337.
Bunge, William W., 1973. "The Geography of Human Survival,"*Annals of the Association of American Geographers*, Vol. 63, pp. 275–295.
Burrill, Meredith F., 1940. "The Printing Industry: A Study in Zonal Agglomeration," *Annals of the Association of American Geographers*, Vol 30, p. 50.
Burrill, Meredith, F., 1956. "East Is North in Montreal," *Professional Geographer*, Vol. 8, pp. 4–5.
Burrill, Meredith, F., 1968. "The Language of Geography," *Annals of the Association of American Geographers*, Vol. 58, pp. 1–12.
Buttimer, Anne, 1976. "Grasping the Dynamism of Lifeworld," *Annals of the Association of American Geographers*, Vol. 66, pp. 277–292.
Butzer, Karl W., 1980. "Adaptation to Global Environmental Change," *Professional Geographer*, Vol. 32, pp. 269–278.
Cahnman, Werner J., 1948. "Outline of a Theory of Area Studies," *Annals of the Association of American Geographers*, Vol. 38, pp. 233–243.
Campbell, M. R., 1932. "A Composite Peneplain," *Annals of the Association of American Geographers*, Vol. 22, pp. 50–51.
Campbell, Marius R., 1928. "Geographic Terminology," *Annals of the Association of American Geographers*, Vol. 18, pp. 25–40.
Campbell, Robert D., 1968. "Personality As an Element of Regional Geography," *Annals of the Association of American Geographers*, Vol. 58, pp. 748–759.
Cannon, W. A., 1914. "Recent Exploration in the Western Sahara," *Bulletin of the American Geographical Society*, Vol. 46, pp. 81–99.
Carlson, F. A., 1930. "Reconstructing the Geography of Alabama," *Journal of Geography*, Vol. 29, pp. 258–265.
Carney, Frank, 1911. "The Value of the Physical vs. the Human Element in Secondary School Geography," *Journal of Geography*, Vol. 10, pp. 1–7.
Carol, Hans, 1960. "Field Training for Graduate Students," *Annals of the Association of American Geography*, Vol. 50, p. 309.
Chandon, Roland E., 1962. "Geography: A Working Definition," *Journal of Geography*, Vol. 61, pp. 71–75.
Chisholm, George G., 1927. "World Unity," *Geographical Review*, Vol. 17, p. 287–300.
Chorley, Richard J., 1964. "Geography and Analogue Theory," *Annals of the Association of American Geographers*, Vol. 54, pp. 127–137.
*Christian Science Monitor*, November 21, 1919.
Church, J. E., 1933. "Snow Surveying: Its Principles and Possibilities," *Geographical Review*, Vol. 23, pp. 529–563.
Clark, Andrew H., 1962. "*Praemia Geographiae*: The Incidental Rewards of a Professional Career," *Annals of the Association of American Geographers*, Vol. 52, pp. 229–241.
Clark, Gordon L., and Michael Dear, 1978. "The Future of Radical Geography," *Professional Geographer*, Vol. 30, pp. 356–360.

Clark, W. A. V., and Ronald R. Boyce, 1962. ''Brain Power as a Resource for Industry,'' *Professional Geographer*, Vol. 14, pp. 14–16.

Clarkson, James D., 1970. ''Ecology and Spatial Analysis,'' *Annals of the Association of American Geographers*, Vol. 60, pp. 700–716.

Close, C. F., 1910. ''The Purpose and Position of Geography,'' *Bulletin of the American Geographical Society*, Vol. 43, pp. 740–753.

Colby, Charles C., 1931. ''Centrifugal and Centripetal Forces in Urban Geography,'' *Annals of the Association of American Geographers*, Vol. 21, pp. 118–120.

Colby, Charles C., 1936. ''Changing Currents of Geographic Thought in America,'' *Annals of the Association of American Geography*, Vol. 26, pp. 1–37.

Coleson, Edward, 1952. ''Teaching Locational Geography on the Elementary Level,'' *Journal of Geography*, Vol. 51, pp. 147–151.

Cooper, C. E., 1948. ''Some Principles of Geography,'' *Journal of Geography*, Vol. 47, pp. 234–239.

Cooper, Sherwin H., 1966. ''Theoretical Geography, Applied Geography, and Planning,'' *Professional Geographer*, Vol. 18, pp. 1–2.

Couclelis, Helen, and Reginald Golledge, 1983. ''Analytic Research, Positivism, and Behavioral Geography,'' *Annals of the Association of American Geographers*, Vol. 73, pp. 331–339.

Coulter, John Wesley, 1927. ''Contrasts in Dairying in Two Dissimilar Areas of the United States,'' *Geographical Review*, Vol. 17, pp. 605–610.

Coulter, John Wesley, 1949. ''The Method of Science in Human Geography,'' *Annals of the Association of American Geographers*, Vol. 39, pp. 60–61.

Cousin, Victor, 1832. *Introduction to the History of Philosophy*, Boston: Hilliard, Gray, Little, and Wilkins, pp. 240–241.

Cowling, Mary Jo, 1929. ''The Relationship of the Coal Fields and the Population of England and Wales,'' *Bulletin of the Geographical Society of Philadelphia*, Vol. 27, pp. 54–63.

Cressey, George B., 1935. ''The Major Geographic Regions of Eurasia,'' *Journal of Geography*, Vol. 34, pp. 297–301.

Crist, Raymond E., 1932. ''Along the LLanos-Andes Border in Zamora, Venezuela,'' *Geographical Review*, Vol. 22, pp. 411–422.

Crist, Raymond E., 1952. ''The Canning of Guava Fruits: An Item in the Industrialization of Cuba,'' *Journal of Geography*, Vol. 51, pp. 338–341.

Crist, Raymond E., 1955. ''Along the Llanos-Andes Border in Venezuela—Then and Now,'' *Annals of the Association of American Geography*, Vol. 45, p. 176.

Crist, Raymond E., 1969. ''Geography,'' *Professional Geographer*, Vol. 21, pp. 305–307.

Cushing, Sumner W., 1913. ''Coastal Plains and Block Mountains in Japan,'' *Annals of the Association of American Geography*, Vol. 3, pp. 43–61.

Curry, Leslie, 1964. ''The Random Spatial Economy: An Exploration in Settlement Theory,'' *Annals of the Association of American Geographers*, Vol. 54, pp. 138–146.

Dacey, Michael F., 1966. ''A Probability Model for Central Place Locations,'' *Annals of the Association of American Geographers*, Vol. 56, pp. 550–568.

Dakin, W. S., 1931. ''Sources of Geographical Information,'' *Journal of Geography*, Vol. 30, pp. 294–296.

Danklefsen, Mildred, 1953. ''Televising Geography,'' *Journal of Geography*, Vol. 52, pp. 253–257.

Davis, W. M., 1899. *The International Geographer*, New York: D. Appleton & Co.

Davis, W. M., 1902. ''Systematic Geography,'' *Proceedings of the American Philosophical Society*, pp. 235–259.

Davis, W. M., 1911. ''An Item for Commercial Geography,'' *Journal of Geography*, Vol. 9, pp. 157–158.

Davis, W. M., 1911. ''Short Studies Abroad—The Seven Hills of Rome,'' *Journal of Geography*, Vol. 9, pp. 230–233.

Davis, W. M., 1914. ''The Home Study of Coral Reefs,'' *Bulletin of the American Geographical Society*, Vol. 46, pp. 561–577.

Davis, W. M., 1915. ''The Principles of Geographic Description,'' *Annals of the Association of American Geography*, Vol. 5, pp. 61–105.

Davis, W. M., 1924. ''The Progress of Geography in the United States,'' *Annals of the Association of American Geographers*, Vol. 14, pp. 159–215.

Davis, W. M., 1932. ''A Retrospect of Geography,'' *Annals of the Association of American Geography*, Vol. 22, pp. 211–230.

Davis, W. M., 1933. ''Remarks by W. M. Davis on Receiving the Distinguished Service Award of the National Council of Geography Teachers,'' *Journal of Geography*, Vol. 32, pp. 91–95.

Davis, W. M., 1934. ''The Long Beach Earthquake,'' *Geographical Review*, Vol. 24, pp. 1–11.

Davis, W. M., 1934. ''Submarine Mock Valleys,'' *Geographical Review*, Vol. 24, pp. 297–308.

Davis, Wayne K., 1966. ''Latent Migration Potential and Space Preferences,'' *Professional Geographer*, Vol. 18, pp. 300–304.

Davis, William M., 1922. ''Home Geography,'' *Journal of Geography*, Vol. 21, pp. 28–32.

Davis, William Morris, 1906. ''An Inductive Study of the Content of Geography, *Bulletin of the American Geographical Society*, Vol. 38, pp. 67–84.

Deasy, George F., 1937. ''Some Effects of Weather on the Urban Dweller,'' *Journal of Geography*, Vol. 36, p. 106.

De Greer, Sten, 1923. ''On the Definition, Method and Classification of Geography,'' *Geografiska Annales*, Vol. 5, pp. 1–37.

De Greer, Sten, 1923. ''Greater Stockholm: A Geographical Interpretation,'' *Geographical Review*, Vol. 13, pp. 497–506.

Delaisi, Francis, 1927. *Political Myths and Economic Realities*, New York: Viking Press, pp. 138–140.

Denevan, William M., 1983. ''Adaptation, Variation, and Cultural Geography,'' *The Professional Geographer*, Vol. 35, pp. 399–406.

Deskins, Donald R., Jr., 1969. ''Geographical Literature on the American Negro, 1949–1968: A Bibliography,'' *Professional Geographer*, Vol. 21, pp. 145–149.

de Terra, Hellmut, 1939. ''The Quaternary Terrace System of Southern Asia and the Age of Man,'' *Geographical Review*, Vol. 29, pp. 101–118.

Dickinson, Robert E., 1934. ''The Metropolitan Regions of the United States,'' *Geographical Review*, Vol. 24, pp. 278–291.

Dickinson, Robert E., 1938. ''The Economic Regions of Germany,'' *Geographical Review*, Vol. 28, pp. 609–626.

Dickson, Billie L., 1931. "The 'Why' of Spokane," *Journal of Geography*, Vol. 30, pp. 151–160.

Dobby, E. H. G., 1940. "Singapore: Town and Country," *Geographical Review*, Vol. 30, pp. 84–108.

Dobby, E. H. G., 1942. "Settlement Patterns in Malaya," *Geographical Review*, Vol. 32, pp. 211–232.

Dobson, Jerome E., 1983. "Automated Geography," *The Professional Geographer*, Vol. 35, pp. 135–143.

Dodge, Richard Elwood, 1910. "Man and His Geographic Environment," *Journal of Geography*, Vol. 8, pp. 179–187.

Dodge, Richard E., 1915. "An Aesthetic Side of Geography—Beauty in Landscape Forms," *Journal of Geography*, Vol. 13, pp. 302–305.

Dodge, Richard Elwood, 1916. "Some Problems in Geographic Education with Special Reference to Secondary Schools," *Annals of the Association of American Geographers*, Vol. 6, pp. 3–18.

Dodge, Stanley, 1932. "Bureau and the Princeton Community," *Annals of the Association of American Geographers*, Vol. 22, pp. 159–209.

Dohrs, Fred E., 1958. "The Measurement of Location in Political Geography," *Annals of the Association of American Geographers*, Vol. 48, pp. 259–260.

Dominian, Leon, 1913. "The Balkan Peninsula," *Bulletin of the American Geographical Society*, Vol. 45, pp. 576–584.

Dopp, Mary, 1913. "Geographical Influences in the Development of Wisconsin," *Bulletin of the American Geographical Society*, Vol. 45, pp. 401–412.

Dornback, John E., 1959. "The Mental Map," *Annals of the Association of American Geographers*, Vol. 49, pp. 179–180.

Douglass, A. E., 1914. "A Method of Estimating Rainfall by the Growth of Trees," *Bulletin of the American Geographical Society*, Vol. 46, pp. 321–335.

Dryer, Charles R., 1901. *Lessons in Physical Geography*, New York: American Book Company.

Dryer, C. R., 1911. "Philosophical Geography," *Annals of the Association of American Geographers*, Vol. 1, p. 107.

Dryer, Charles R., 1911. "Field Work in Physical Geography," *Journal of Geography*, Vol. 10, pp. 8–12.

Dryer, Charles Redway, 1912. "Regional Geography," *Journal of Geography*, Vol. 11, pp. 73–75.

Dryer, Charles R., 1913. "The New Departure in Geography," *Journal of Geography*, Vol. 11, pp. 145–151.

Dryer, Charles R., 1915. "Natural Economic Regions," *Annals of the Association of American Geographers*, Vol. 5, pp. 121–125.

Dryer, Charles Redway, 1920. "Genetic Geography: The Development of the Geographic Sense and Concept," *Annals of the Association of American Geographers*, Vol. 10, pp. 3–16.

Dunbar, G. S., 1978. "What *Was* Applied Geography?" *Professional Geographer*, Vol. 30, pp. 238–239.

Durand, Loyal, Jr., 1932. "The Geographic Regions of Wisconsin," *Annals of the Association of American Geographers*, Vol. 22, pp. 55–56.

Durand, Loyal, Jr., 1941. "The Glaciated Sand Region of Central Wisconsin," *Annals of the Association of American Geographers*, Vol. 31, pp. 289–309.

Dury, G. H., 1959. *The Face of the Earth*, London, Penguin.

Dury, George Harry, 1972. "Some Recent Views on the Nature, Location, Needs, and Potential of Geomorphology," *Professional Geographer*, Vol. 24, pp. 199–202.

Dutton, C. E., 1880–1881. "The Physical Geology of the Grand Cañon District," *Second Annual Report of the U.S. Geologic Survey, 1880–1881*. Washington, D.C., p. 51.

Eiselen, Elizabeth, 1938. "Celery Growing in the United States," *Journal of Geography*, Vol. 37, pp. 32–36.

Ekblaw, W. Elmer, 1931. "Russia Today," *Annals of the Association of American Geographers*, Vol. 21, pp. 121–122.

Ekblaw, W. Elmer, 1937. "The Attributes of Place," *Journal of Geography*, Vol. 36, pp. 213–220.

Ekblaw, W. Elmer, 1937. "The Significance of Soils Geography," *Annals of the Association of American Geographers*, Vol. 27, pp. 103–104.

Emerson, F. V., 1911. "Geographic Influences in American Slavery," *Bulletin of the American Geographical Society*, Vol. 43, pp. 13–26.

Emerson, Philip, 1911. "General Truths in Geography as Illustrated by New England Relief and Life," *Journal of Geography*, Vol. 10, pp. 20–23.

Entrikin, J. Nichols, 1976. "Contemporary Humanism in Geography," *Annals of the Association of American Geographers*, Vol. 66, pp. 615–632.

Estes, John E., 1966. "Some Applications of Aerial Infrared Imagery," *Annals of the Association of American Geographers*, Vol. 56, pp. 673–683.

Ezequiel, Martinez Estrada, 1971. *X-Ray of the Pampa*, Austin, Texas: University of Texas Press, p. 143.

Fassig, Oliver L., 1916. "A Simplified Form of Revolving Cloud Camera," *Annals of the Association of American Geographers*, Vol. 6, p. 125.

Fawcett, C. B., 1931. "England," *Journal of Geography*, Vol. 30, pp. 111–119.

Fead, Margaret Irene, 1932. "The Development of the Cartographical Representation of Cities," *Annals of the Association of American Geographers*, Vol. 22, p. 56.

Fead, Margaret Irene, 1933. "Notes on the Development of the Cartographic Representation of Cities," *Geographical Review*, Vol. 23, pp. 441–456.

Fenneman, N. M., 1911. "Geographic Influences Affecting Early Cincinnati," *Journal of Geography*, Vol. 9, p. 192.

Fenneman, Nevin M., 1916. "Physiographic Divisions of the United States," *Annals of the Association of American Geographers*, Vol. 6, pp. 19–98.

Fenneman, Nevin M., 1919. "The Circumference of Geography," *Annals of the Association of American Geographers*, Vol. 9, pp. 3–11.

Fenneman, Nevin M., 1920. "Geography as a Subject of Research," *Annals of the Association of American Geographers*, Vol. 10, p. 154.

Fenneman, Nevin M., 1922. "The Place of Physiography in Geography," *Journal of Geography*, Vol. 21, pp. 20–23.

Fewkes, Walter, 1914. "Relations of Aboriginal Culture and Environment in the Lesser Antilles," *Bulletin of the American Geograhical Society*, Vol. 46, pp. 662–678.

Fifield, Russell H., 1944. "The Geostrategy of Location," *Journal of Geography*, Vol. 43, pp. 297–303.

Finch, V. C., 1930. "An Introductory Course in College Geography for Liberal Arts Students," *Journal of Geography*, Vol. 29, pp. 178–186.

Finch, V. C., 1934. "Written Structures for Presenting the Geography of Regions," *Annals of the Association of American Geographers*, Vol. 24, pp. 113–120.

Finch, V. C., 1939. "Geographical Science and Social Philosophy," *Annals of the Association of American Geographers*, Vol. 29, pp. 1–28.

Finley, Virginia P., 1958. "Aerial Photographs as Basic Source Material in Geographic Research: An Example from the Grand Coulee Area," *Annals of the Association of American Geographers*, Vol. 48, p. 261.

Fleming, Douglas K., 1984. "Cartographic Strategies for Airline Advertising," *Geographical Review*, Vol. 74, pp. 76–93.

Floyd, Barry N., 1962. "The Pleasures Ahead: A Geographic Meditation," *Professional Geographer*, Vol. 14, pp. 1–4.

Floyd, Barry, 1963. "Quantification—A Geographic Deviation?" *Professional Geographer*, Vol. 15, pp. 15–21.

Forbes, Robert H., 1933. "The Black Man's Industries," *Geographical Review*, Vol. 23, pp. 230–247.

Ford, Larry R., 1982. "Beware of New Geographies," *Professional Geographer*, Vol. 34, pp. 131–135.

Ford, Larry, and Ernst Griffin, 1979. "The Ghettoization of Paradise," *Geographical Review*, Vol. 69, pp. 140–158.

Fosberg, F. R., 1976. "Geography, Ecology, and Biogeography," *Annals of the Association of American Geographers*, Vol. 66, pp. 117–128.

Francaviglia, Richard V., 1971. "The Cemetery As an Evolving Cultural Landscape," *Annals of the Association of American Geographers*, Vol. 61, pp. 501–509.

Frazier, John W., 1978. "On the Emergence of an Applied Geography," *Professional Geographer*, Vol. 30, pp. 233–237.

Freeman, Otis W., 1936. "Teaching Geographic Relations in World Problems," *Journal of Geography*, Vol. 35, pp. 90–98.

Frey, John W., 1933. "A Few Notes on the Geography of the American Oil Industry," *Annals of the Association of American Geographers*, Vol. 23, pp. 44–45.

Frey, John W., 1940. "The World's Petroleum," *Geographical Review*, Vol. 30, pp. 451–462.

Fuson, Robert H., Harry J. Schaleman, Jr., and Douglas C. Wilms, 1978. "Baseball and Geography: The Hidden Connection," *Professional Geographer*, Vol. 30, pp. 319–321.

Gade, Daniel W., 1983. "Foreign Languages and American Geography," *The Professional Geographer*, Vol. 35, pp. 261–266.

Gale, Stephen, 1977. "Ideological Man in a Nonideological Society," *Annals of the Association of American Geographers*, Vol. 67, pp. 267–272.

Gardner, Alan S., 1954. "Applications of Geography in Marketing," *Annals of the Association of American Geographers*, Vol. 44, p. 208.

Garnett, Alice, 1935. "Insolation, Topography, and Settlement in the Alps," *Geographical Review*, Vol. 25, pp. 601–617.

Garnier, B. J., 1963. "A Program for Physical Geography," *Professional Geographer*, Vol. 15, pp. 16–18.

Garrison, William L., 1956. "Some Confusing Aspects of Common Measures," *Professional Geographer*, Vol. 8, pp. 4–5.

Garrison, William L., 1979. "Playing with Ideas," *Annals of the Association of American Geographers*, Vol. 69, pp. 118–120.

Garrison, William L., and Duane F. Marble, 1957. ''The Spatial Structure of Agricultural Activities,'' *Annals of the Association of American Geographers*, Vol. 47, pp. 136–144.

Gast, Marvin, 1962. ''The Overhead Projector,'' *Annals of the Association of American Geographers*, Vol. 52, p. 333.

Gentha, Martha Krug, 1913. ''Notes on the History of Gotha Cartography,'' *Bulletin of the American Geographical Society*, Vol. 45, pp. 33–38.

Gerlack, Arch C., 1958. ''Linguistic Maps,'' *Annals of the Association of American Geographers*, Vol. 48, p. 262.

Gersmehl, Philip J., 1976. ''An Alternative Biogeography,'' *Annals of the Association of American Geographers*, Vol. 66, pp. 223–241.

Gibson, Lyle E., 1948. ''Characteristics of a Regional Margin of the Corn and Dairy Belts,'' *Annals of the Association of American Geographers*, Vol. 38, pp. 244–270.

Gildea, Ray Y., Jr., 1957. ''Synthesis of Geographic, Economic and Legal Disciplines in the Analysis of Water Management Problems,'' *Journal of Geography*, Vol. 56, pp. 433–437.

Gilmore, Melvin R., 1927. ''The Missouri River and the Indians,'' *Bulletin of the Geographical Society of Philadelphia*, Vol. 25, pp. 155–161.

Ginsburg, Norton, 1972. ''The Mission of a Scholarly Society,'' *Professional Geographer*, Vol. 24, pp. 1–6.

Ginsburg, Norton, 1973. ''From Colonialism to National Development: Geographical Perspectives on Patterns and Policies,'' *Annals of the Association of American Geographers*, Vol. 63, pp. 1–21.

Glendinning, Robert M., 1940. ''The Role of Death Valley,'' *Annals of the Association of American Geographers*, Vol. 30, p. 56.

Golledge, Reginald G., and Douglas M. Amedeo, 1966. ''Some Introductory Notes on Regional Division and Set Theory,'' *Professional Geographer*, Vol. 18, pp. 14–19.

Golledge, Reginald G., and Douglas Amedeo, 1968. ''On Laws in Geography,'' *Annals of the Association of American Geographers*, Vol. 58, pp. 760–774.

Goode, J. Paul, 1911. ''The Point of View in Elementary Commercial Geography,'' *Journal of Geography*, Vol. 10, pp. 155–156.

Goode, J. Paul, 1924. ''The Evil Mercator,'' *Annals of the Association of American Geographers*, Vol. 14, p. 39.

Goode, J. Paul, 1927. ''The Map as a Record of Progress in Geography,'' *Annals of the Association of American Geographers*, Vol. 17, pp. 1–14.

Goode, Nat J., Jr., 1968. ''The Professional Geographer's Contribution to the Retail Food Industry,'' *Professional Geographer*, Vol. 20, pp. 396–397.

Goodrich, L. Carrington, 1938. ''China's First Knowledge of the Americas,'' *Geographical Review*, Vol. 28, pp. 400–411.

Gorgas, W. C., 1915. *Boston Medical and Surgical Journal*, p. 220.

Gottmann, Jean, 1937. ''The Pioneer Fringe in Palestine,'' *Geographical Review*, Vol. 27, pp. 550–565.

Gould, Peter, 1981. ''Letting the Data Speak for Themselves,'' *Annals of the Association of American Geographers*, Vol. 71, pp. 166–176.

Graf, William L., et al., 1980. ''Geographic Geomorphology in the Eighties,'' *Professional Geography*, Vol. 32, pp. 279–284.

Grigg, David, 1965. ''The Logic of Regional Systems,'' *Annals of the Association of American Geographers*, Vol. 55, pp. 465–491.

Griswold, Erwin N., 1939. ''Hunting Boundaries with Car and Camera in the Northeastern United States,'' *Geographical Review*, Vol. 29, pp. 353–382.

Grosvenor, Gilbert M., 1984. ''The Society and the Discipline,'' *Professional Geographer*, Vol. 36, pp. 413–418.

Grvaldy-Sczny, W. T., 1979. ''A Diamond Anniversary,'' *Annals of the Association of American Geographers*, Vol. 69, pp. 1–3.

Guelke, Leonard, 1974. ''An Idealist Alternative in Human Geography,'' *Annals of the Association of American Geographers*, Vol. 64, pp. 193–202.

Guelke, Leonard, 1977. ''Regional Geography,'' *Professional Geographer*, Vol. 29, p. 107.

Haas, W. H., 1931. ''The Teaching of Geography as a Science,'' *Journal of Geography*, Vol. 30, pp. 323–329.

Haas, William H., 1934. ''Foundations and Limits in Geography,'' *Annals of the Association of American Geographers*, Vol. 24, p. 53.

Haggett, Peter, 1966. *Locational Analysis in Human Geography*, New York: St. Martin's Press, p. 277.

Hall, Robert Burnett, 1935. ''The Geographic Region: A Resume,'' *Annals of the Association of American Geographers*, Vol. 25, pp. 122–130.

Halterberger, Michael, 1915. ''Primitive Carriers in Land Transportation,'' *Bulletin of the American Geographical Society*, Vol. 47, pp. 729–745.

Halverson, L. H., 1930. ''The Great Karroo of South Africa,'' *Journal of Geography*, Vol. 29, pp. 287–300.

Hammond, Edwin H., 1964. ''Analysis of Properties in Land Form Geography: An Application to Broad-Scale Land Form Mapping,''*Annals of the Association of American Geographers*, Vol 54, pp. 11–19.

Hardy, Osgood, 1914. ''Cuzco and Apurimac,'' *Bulletin of the American Geographical Society*, Vol. 46, pp. 500–512.

Hare, F. Kenneth, 1955. ''Dynamic and Synoptic Climatology,'' *Annals of the Association of American Geographers*, Vol. 45, pp. 152–162.

Hare, F. Kenneth, 1964. ''New Light from Labrador Ungava,'' *Annals of the Association of American Geographers*, Vol. 54, pp. 459–476.

Harris, Chauncy D., 1942. ''Growth of the Larger Cities in the United States, 1930–1940,'' *Journal of Geography*, Vol. 41, pp. 313–318.

Harris, Chauncy D., 1954. ''The Market as a Factor in the Localization of Industry in the United States,'' *Annals of the Association of American Geographers*, Vol. 44, pp. 315–348.

Harris, Chauncy D., 1958. ''Geography in the Soviet Union,'' *Professional Geographer*, Vol. 10, pp. 8–13.

Hart, Isabelle K., 1931. ''The Need for and Nature of Journey Geography in the Fourth Grade,'' *Journal of Geography*, Vol. 30, pp. 170–176.

Hart, John Fraser, 1957. ''Value Systems and Geographic Research,'' *Annals of the Association of American Geographers*, Vol. 47, p. 164.

Hart, John Fraser, 1968. ''Loss and Abandonment of Cleared Farm Land in the Eastern United States,'' *Annals of the Association of American Geographers*, Vol. 58, pp. 417–440.

Hart, John Fraser, 1982. "The Highest Form of the Geographer's Art," *Annals of the Association of American Geographers*, Vol. 72, pp. 1–29.

Hartshorne, Richard, 1927. "Location as a Factor in Geography," *Annals of the Association of American Geographers*, Vol. 17, pp. 92–99.

Hartshorne, Richard, 1932. "The Twin City District: A Unique Form of Urban Landscape," *Geographical Review*, Vol. 22, pp. 431–442.

Hartshorne, Richard, 1933. "Geographic and Political Boundaries in Upper Silesia," *Annals of the Association of American Geographers*, Vol. 23, pp. 195–228.

Hartshorne, Richard, 1937. "The Polish Corridor," *Journal of Geography*, Vol. 36, pp. 161–176.

Hartshorne, Richard, 1938. "Racial Maps of the United States," *Geographical Review*, Vol. 28, pp. 276–288.

Hartshorne, Richard, 1939. "The Nature of Geography: A Critical Survey of Current Thought in the Light of the Past," *Annals of the Association of American Geographers*, Vol. 29, pp. 177–658.

Hartshorne, Richard, 1948. "On the Mores of Methodological Discussion in American Geography," *Annals of the Association of American Geographers*, Vol. 38, pp. 113–125.

Hartshorne, Richard, 1953. "Where in the World Are We? Geographic Understanding for Political Survival and Progress," *Journal of Geography*, Vol. 52, pp. 382–393.

Hartshorne, Richard, 1955. " 'Exceptionalism in Geography' Re-Examined," *Annals of the Association of American Geographers*, Vol. 45, pp. 206–244.

Hartshorne, Richard, 1958. "The Concept of Geography as a Science of Space, from Kant and Humboldt to Hettner," *Annals of the Association of American Geographers*, Vol. 48, pp. 97–108.

Hartshorne, Richard, 1958. "What Do We Mean by 'Region'?" *Annals of the Association of American Geographers*, Vol. 48, p. 268.

Hartshorne, Richard, 1962. "On the Concept of Areal Differentiation," *Professional Geographer*, Vol. 14, pp. 10–12.

Hartshorne, Richard, 1965. "Practical and Academic Regions," *Annals of the Association of American Geographers*, Vol. 55, p. 619.

Hartz, Robert E., 1931. "Map Reading by Aviators," *Journal of Geography*, Vol. 30, p. 339.

Harvey, David, 1979. "Monument and Myth," *Annals of the Association of American Geographers*, Vol. 69, pp. 362–381.

Harvey, David, 1984. "On the History and Present Condition of Geography: An Historical Materialist Manifesto," *The Professional Geographer*, Vol. 36, pp. 1–11.

Hasson, Shlomo, 1984. "Humanistic Geography from the Perspective of Martin Buber's Philosophy," *Professional Geographer*, Vol. 36, pp. 11–18.

Hay, Francis S., 1925. "Sheep and Human Affairs," *Journal of Geography*, Vol. 24, pp. 350–357.

Helburn, Nicholas, 1957. "The Basis for a Classification of World Agriculture," *Professional Geographer*, Vol. 9, pp. 2–7.

Helburn, Nicholas, 1977. "The Wilderness Continuum," *Professional Geographer*, Vol. 29, pp. 333–337.

Helburn, Nicholas, 1982. "Geography and the Quality of Life," *Annals of the Association of American Geographers*, Vol. 72, pp. 445–456.

Henderson, Bertha, 1910. ''The Modern Trend of Geography'', *Journal of Geography*, Vol. 8, pp. 129–134.

Herbert, John W., 1913. ''The Panama Canal: Its Construction and Its Effect on Commerce,'' *Bulletin of the American Geographical Society*, Vol. 45, pp. 241–254.

Hettner, Alfred, 1927. *Die Geographie*, Breslan.

Hewings, Geoffrey J. D., 1984. ''New Directions in Regional and Interregional Modelling: Introduction,'' *Economic Geography*, Vol. 60, pp. 99–101.

Hill, A. David, 1982. ''Another View of the Sixteen-Million-Hour Question,'' *The Professional Geographer*, Vol. 34, pp. 1–5.

Hodgkins, Alton R., 1925. ''The Study of Economic Geography,'' *Journal of Geography*, Vol. 24, pp. 226–234.

Holdridge, Desmond, 1940. ''Toledo: A Tropical Refugee Settlement in British Honduras,'' *Geographical Review*, Vol. 30, pp. 376–393.

Holway, Ruliff S., 1911. ''Topographic Environment of San Francisco,'' *Journal of Geography*, Vol. 9, pp. 261–265.

Holzner, Lutz, 1966. ''Urban Geography Without a Regional System,'' *Professional Geographer*, Vol. 18, pp. 129–131.

Hoosen, David J. M., 1958. ''Some Recent Developments in the Content and Theory of Soviet Geography,'' *Annals of the Association of American Geographers*, Vol. 48, pp. 269–270.

Hoover, J. W., 1931. ''Geographic and Ethnic Grouping of Arizona Indians,'' *Journal of Geography*, Vol. 30, pp. 235–246.

Horvath, Ronald J., Donald R. Deskins, Jr., and Ann E. Larimore, 1969. ''Activity Concerning Granting M.A. and Ph.D. Degrees in Geography,'' *Professional Geographer*, Vol. 21, pp. 137–139.

Howland, Felix, 1940. ''Crossing the Hindu Kush,'' *Geographical Review*, Vol. 30, pp. 272–278.

Hsieh, Chiao-Min, 1959. ''Geography in Communist China,'' *Annals of the Association of American Geographers*, Vol. 49, pp. 186–187.

Hubbard, George D., 1940. ''Major Objectives of Penck and of Davis in Geomorphic Studies,'' *Annals of the Association of American Geographers*, Vol. 30, p. 60.

Hudson, G. Donald, 1935. ''Geography and Regional Planning,'' *Journal of Geography*, Vol. 34, pp. 267–277.

Hudson G. Donald, 1936. ''The Unit Area Method of Land Classification,'' *Annals of the Association of American Geographers*, Vol. 26, pp. 99–112.

Hudson, G. Donald, 1951. ''Professional Training of the Membership of the Association of American Geographers,'' *Annals of the Association of American Geographers*, Vol. 41, pp. 97–115.

Hung, F., 1966. ''Comments on *Frontiers in Geographical Teaching*,'' *Professional Geographer*, Vol. 18, pp. 341–345.

Hunter, James M., 1958. ''Geodiplomacy and the Eisenhower Doctrine,'' *Annals of the Association of American Geographers*, Vol. 48, p. 271.

Hunter, James M., 1960. ''Examination of a Concept Basic in Ratzel's Methodology of Political Geography: The State Is a Spiritual Moral Organism,'' *Annals of the Association of American Geographers*, Vol. 50, p. 327.

Huntington, Ellsworth, 1912. ''The Effect of Barometric Variations upon Mental Activities,'' *Annals of the Association of American Geographers*, Vol. 2, p. 116.

Huntington, Ellsworth, 1913. "The New Science of Geography," *Bulletin of the American Geographical Society*, Vol. 45, pp. 641–652.
Huntington, Ellsworth, 1913. "The Shifting of Climatic Zones as Illustrated in Mexico," *Bulletin of the American Geographical Society*, Vol. 45, pp. 107–116.
Huntington, Ellsworth, 1916. "Weather and Civilization," *Bulletin of the Geographical Society of Philadelphia*, Vol. 14, pp. 1–21.
Huntington, Ellsworth, 1922. "Influenza, an Example of Statistical Geography," *Annals of the Association of American Geographers*, Vol. 12, pp. 210–211.
Huntington, Ellsworth, 1924. "Geography and Natural Selection: A Preliminary Study of the Origin and Development of Racial Character," *Annals of the Association of American Geographers*, Vol. 14, pp. 1–16.
Huntington, Ellsworth, 1926. "The Handicap of Poor Land," *Economic Geography*, Vol. 2, pp. 335–357.
Huntington, Ellsworth, 1932. "The Measurement and Geographical Distribution of Mental Activity," *Annals of the Association of American Geographers*, Vol. 22, pp. 62–63.
Huntington, Ellsworth, 1935. "Caravan Cities, Climate, and History," *Annals of the Association of American Geographers*, Vol. 25, p. 45.
Huntington, Ellsworth, 1937. "Season of Birth, and the Distribution of Civilization," *Annals of the Association of American Geographers*, Vol. 27, pp. 109–110.
Huntington, Ellsworth, 1938. "Farms and Villages of Sweden," *Journal of Geography*, Vol. 37, pp. 85–90.
Huntington, Ellsworth, 1942. "What Next in Geography?" *Journal of Geography*,Vol. 41, pp. 1–9.
Huntington, Ellsworth, 1943. "The Geography of Human Productivity," *Annals of the Association of American Geographers*, Vol. 33, pp. 1–31.
Huntington, Ellsworth, 1945. "High Schools and Geographic Immaturity," *Journal of Geography*, Vol. 44, pp. 173–181.
Huntington, Ellsworth, and Sumner W. Cushing, 1919. "The Nature and Possibilities of Tropical Agriculture," *Journal of Geography*, Vol. 18, pp. 341–352.
Huntington, Ellsworth, and Sumner W. Cushing, 1920. "The Rivalry Between Sugar Beets and Sugar Cane," *Journal of Geography*, Vol. 19, pp. 255–259.
Jackson, Eric P., 1924. "Geography as a Correlating Subject in the High School," *Journal of Geography*, Vol. 23, pp. 313–316.
Jackson, J. B., 1972. "Metermorphosis," *Annals of the Association of American Geographers*, Vol. 62, pp. 155–158.
Jackson, W. A. Douglas, 1958. "Whither Political Geography?" *Annals of the Association of American Geographers*, Vol. 48, pp. 178–183.
James, Preston E., 1925. "The Pitch Lake, Trinidad," *Journal of Geography*, Vol. 24, pp. 212–220.
James, Preston E., 1929. "The Blackstone Valley: A Study in Chorography in Southern New England," *Annals of the Association of American Geographers*, Vol. 19, pp. 67–109.
James, Preston E., 1930. "The Shari Plain," *Journal of Geography*, Vol. 29, pp. 319–330.
James, Preston E., 1934. "The Terminology of Regional Description," *Annals of the Association of American Geographers*, Vol. 24, pp. 78–92.

James, Preston E., 1939. "Air Masses and Fronts in South America," *Geographical Journal*, Vol. 29, pp. 132–134.

James, Preston E., 1946. "Differences from Place to Place," *Journal of Geography*, Vol. 45, pp. 279–285.

James, Preston E., 1947. "Developments in the Field of Geography and Their Implications for the Geography Curriculum," *Journal of Geography*, Vol. 46, pp. 221–226.

James, Preston E., 1948. "Formulating Objectives of Geographic Research," *Annals of the Association of American Geographers*, Vol. 38, pp. 271–276.

James, Preston E., 1952. "Toward a Further Understanding of the Regional Concept," *Annals of the Association of American Geographers*, Vol. 42, pp. 195–222.

James, Preston E., 1963. "Geography in an Age of Revolution," *Journal of Geography*, Vol. 62, pp. 98–103.

James, Preston E., 1967. "On the Origin and Persistence of Error in Geography," *Annals of the Association of American Geographers*, Vol. 57, pp. 1–24.

James, Preston E., 1976. "The Process of Competitive Discussion," *Professional Geographer*, Vol. 28, pp. 1–7.

James, Preston E., and R. Burnett Hall, 1924. "A Suggested Outline for the Treatment of a Geographic Region," *Journal of Geography*, Vol. 23, pp. 288–291.

Janelle, Donald G., 1968. "Central Place Development in a Time-Space Framework," *Professional Geographer*, Vol. 20, pp. 5–10.

Jefferson, Mark, 1912. "Houses and House Materials in West Europe," *Annals of the Association of American Geographers*, Vol. 2, p. 106.

Jefferson, Mark, 1913. "The Anthropology of North America," *Bulletin of the American Geographical Society*, Vol. 45, pp. 116–180.

Jefferson, Mark, 1914. "Population Estimates for the Countries of the World from 1914 to 1920," *Bulletin of the American Geographical Society*, Vol. 46, pp. 401–413.

Jefferson, Mark, 1915. "Regional Characters in the Growth of American Cities," *Annals of the Association of American Geographers*, Vol. 5, p. 140.

Jefferson, Mark, 1915. "How American Cities Grow," *Bulletin of the American Geographical Society*, Vol. 47, pp. 19–37.

Jefferson, Mark, 1917. "Some Considerations on the Geographical Provinces of the United States," *Annals of the Association of American Geographers*, Vol. 7, pp. 3–15.

Jefferson, Mark, 1925. "Malthus in Light of Subsequent Events," *Annals of the Association of American Geographers*, Vol. 15, p. 39.

Jefferson, Mark, 1927. "The Sea Island Coast," *Annals of the Association of American Geographers*, Vol. 17, pp. 31–32.

Jefferson, Mark, 1928. "The Civilizing Rails," *Economic Geography*, Vol. 4, pp. 217–231.

Jefferson, Mark, 1929. "The Geographic Distribution of Inventiveness," *Geographical Review*, Vol. 19, pp. 649–661.

Jefferson, Mark, 1931. "The Distribution of Urbanization," *Annals of the Association of American Geographers*, Vol. 21, pp. 126–127.

Jefferson, Mark, 1932. "An Adventure in Geography," *Journal of Geography*, Vol. 30, pp. 201–204.

Jefferson, Mark, 1933. "Communications and Civilization," *Annals of the Association of American Geographers*, pp. 47–48.

Jefferson, Mark, 1933. "Great Cities of 1930 in the United States with a Comparison of New York and London," *Geographical Review*, Vol. 23, pp. 90–100.

Jefferson, Mark, 1934. "Is the Population of the United States Now Decreasing?" *Geographical Review*, Vol. 24, pp. 634–637.

Jefferson, Mark, 1939. "The Law of Primate Cities," *Annals of the Association of American Geographers*, Vol. 29, pp. 78–79.

Jefferson, Mark, 1939. "The Law of the Primate City," *Geographical Review*, Vol. 29, pp. 226–232.

Jefferson, Mark, 1940. "Is Germany a Country? What Is the Test?" *Annals of the Association of American Geographers*, Vol. 30, pp. 60–61.

Jefferson, Mark, 1941. "The Great Cities of the United States, 1940," *Geographical Review*, Vol. 32, pp. 479–487.

Jenks, George F., 1963. "How to Improve Illustrations in American Geographic Journals," *Annals of the Association of American Geographers*, Vol. 53, pp. 599–600.

Jensen, J. Granville, 1945. "Community Vision thru Geography," *Journal of Geography*, Vol. 44, pp. 162–166.

Jett, Stephen C., 1970. "The Development and Distribution of the Blowgun," *Annals of the Association of American Geographers*, Vol. 60, pp. 662–688.

Joerg, Wolfgang L. G., 1914. "The Subdivisions of North America into Natural Regions: A Preliminary Inquiry," *Annals of the Association of American Geographers*, Vol. 4, pp. 55–83.

Joerg, W. L. G., 1923. "The Use of Airplane Photography in City Geography," *Annals of the Association of American Geographers*, Vol. 13, p. 211.

Joerg, W. L. G., 1935. "Geography and National Land Planning," *Geographical Review*, Vol. 25, pp. 177–208.

Joerg, W. L. G., 1936. "The Geography of North America: A History of Its Regional Exposition," *Geographical Review*, Vol. 26, pp. 640–663.

Johnson, Douglas Wilson, 1910. "Map Drawing in the Schools," *Journal of Geography*, Vol. 8, pp. 152–158.

Johnson, Emory R., 1910. "Sources of American Railway Freight Traffic," *Bulletin of the American Geographical Society*, Vol. 42, pp. 243–260.

Johnson, Hildegard Binder, 1956. "Spatial Feeling in Geography Teaching," *Journal of Geography*, Vol. 55, pp. 334–341.

Johnson, Wallis E., 1911. "To Set Your Watch by the Sun," *Journal of Geography*, Vol. 10, pp. 68–70.

Johnston, R. J., 1968. "Choice in Classification: The Subjectivity of Objective Methods," *Annals of the Association of American Geographers*, Vol. 58, pp. 575–589.

Johnston, R. J., 1981. "Applied Geography, Quantitative Analysis and Ideology," *Applied Geography*, Vol. 1, pp. 213–219.

Jones, Clarence F., 1927. "Chilean Commerce," *Economic Geography*, Vol. 3, pp. 139–166.

Jones, Emrys, 1956. "Cause and Effect in Human Geography," *Annals of the Association of American Geographers*, Vol. 46, pp. 369–377.

Jones, Stephen B., 1932. "The Forty-ninth Parallel in the Great Plains: The Historical Geography of a Boundary," *Journal of Geography*, Vol. 31, pp. 357–368.

Jones, Stephen B., 1943. "The Description of International Boundaries," *Annals of the Association of American Geographers*, Vol. 33, pp. 99–117.

Jones, Stephen B., 1954. "A Unified Field Theory of Political Geography," *Annals of the Association of American Geographers*, Vol. 44, pp. 111–123.

Jones, Stephen B., 1959. "Boundary Concepts in the Setting of Place and Time," *Annals of the Association of American Geographers*, Vol. 49, pp. 241–255.

Jones, Wellington D., 1931. "Field Mapping of Residential Areas in Metropolitan Chicago," *Annals of the Association of American Geographers*, Vol. 21, pp. 207–214.

Jones, Wellington D., 1934. "Procedures in Investigating Human Occupance of a Region," *Annals of the Association of American Geographers*, Vol. 24, pp. 93–111.

Jones, Wellington D., and V. C. Finch, 1925. "Detailed Field Mapping in the Study of the Economic Geography of an Agricultural Area," *Annals of the Association of American Geographers*, Vol. 15, pp. 148–157.

Jones, Wellington D., and Carl O. Sauer, 1915. "Outline for Field Work in Geography," *Bulletin of the American Geographical Society*, Vol. 47, pp. 520–525.

Jordan, Terry G., 1966. "On the Nature of Settlement Geography," *Professional Geographer*, Vol. 18, pp. 26–28.

Kalesnik, S. V., 1964. "General Geographic Regularities of the Earth," *Annals of the Association of American Geographers*, Vol. 54, pp. 160–164.

Kansky, Karel J., 1964. "Highway Networks in Southern Illinois and Eastern North Carolina," *Annals of the Association of American Geographers*, Vol. 54, p. 426.

Karpinski, Louis, 1923. "The Contribution of Mathematics and Astronomers to Scientific Cartography," *Annals of the Association of American Geographers*, Vol. 13, pp. 211–212.

Kearns, Kevin C., 1968. "On the Nature and Origin of Parks in Urban Areas," *Professional Geographer*, Vol. 20, pp. 167–170.

Keinard, Marguerite, 1953. "Using Music to Enrich Geography," *Journal of Geography*, Vol. 52, pp. 189–191.

Keir, R. Malcolm, 1914. "Modern Korea," *Bulletin of the American Geographical Society*, Vol. 46, pp. 756–769.

Kellogg, Charles E., 1937. "Soil and People," *Annals of the Association of American Geographers*, Vol. 27, pp. 142–148.

Kemp, Harold S., 1930. "Brittany: The Backward Child of a Stern Environment," *Journal of Geography*, Vol. 29, pp. 223–240.

Kemp, Harold S., 1940. "Mussolini 'Geographer,' " *Annals of the Association of American Geographers*, Vol. 30, pp. 61–62.

Kemp, Harold S., 1940. "Mussolini: Italy's Geographer-in-Chief," *Journal of Geography*, Vol. 39, pp. 133–141.

Kennamer, Lorren G., Jr., 1955. "The Unique Change in School Geography," *Journal of Geography*, Vol. 54, pp. 25–32.

Kimble, George H. T., 1951. "The Role of Geography at the Mid-Century," *Journal of Geography*, Vol. 50, pp. 45–54.

Kimble, G. H. T., 1952. "Expanding Horizons in a Shrinking World," *Geographical Review*, Vol. 42, pp. 1–6.

Kish, George, 1963. "New Directions in Academic Geography," *Annals of the Association of American Geographers*, Vol. 53, p. 602.

Kish, George, 1964. ''Mirror for the American Geographer,'' *Professional Geographer*, Vol. 16, pp. 1–4.

Kiss, George, 1942. ''Political Geography into Geopolitics,'' *Geographical Review*, Vol. 32, pp. 632–645.

Kniffen, Fred B., 1936. ''Louisiana House Types,'' *Annals of the Association of American Geographers*, Vol. 26, pp. 179–193.

Kniffen, Fred B., 1937. ''The Lower Mississippi,'' *Annals of the Association of American Geographers*, Vol. 27, pp. 162–167.

Kniffen, Fred, 1954. ''Whither Cultural Geography,'' *Annals of the Association of American Geographers*, Vol. 44, pp. 222–223.

Kniffen, Fred, 1965. ''Folk Housing: Key to Diffusion,'' *Annals of the Association of American Geographers*, Vol. 55, pp. 549–577.

Koeppe, Clarence E., 1957. ''Proposal for a New Classification of Climates,'' *Professional Geographer*, Vol. 9, pp. 6–8.

Kohn, Clyde F., 1959. ''Spatial Dimensions of Human Activities: Significance for Geographic Education,'' *Journal of Geography*, Vol. 58, pp. 121–127.

Kohn, Clyde F., 1970. ''The 1960s: A Decade of Progress in Geographical Research and Instruction,'' *Annals of the Association of American Geographers*, Vol. 60, pp. 211–219.

Kollmorgen, Walter M., 1969. ''The Woodsman's Assaults on the Domain of the Cattleman,'' *Annals of the Association of American Geographers*, Vol. 59, pp. 215–239.

Kopf, Helen M., 1932. ''One-Cycle or Two-Cycle Geography,'' *Journal of Geography*, Vol. 31, pp. 193–198.

Kosiński, Leszek, 1971. ''Geography of Population and Settlement in East-Central Europe,'' *Annals of the Association of American Geographers*, Vol. 61, pp. 599–615.

Kriesel, Karl Marcus, 1968. ''Montesquieu: Possibilistic Political Geographer,'' *Annals of the Association of American Geographers*, Vol. 58, pp. 557–574.

Kristof, Ladis K. D., 1959. ''The Nature of Frontiers and Boundaries,'' *Annals of the Association of American Geographers*, Vol. 49, pp. 269–282.

Lackey, E. E., 1924. ''The Classification and Use of Geographic Principles,'' *Journal of Geography*, Vol. 23, pp. 67–71.

Laity, Alan L., 1984. ''Perceiving Regions as Scattered Objects,'' *Professional Geographer*, Vol. 36, pp. 285–292.

Latham, Brice M., 1937. ''How Berry Gully was Conquered,'' *Journal of Geography*, Vol. 36, pp. 193–196.

Latham, James P., 1962. ''Methodology for Instrumented Geographic Research,'' *Annals of the Association of American Geographers*, Vol. 52, pp. 346–347.

Lattimore, Owen, 1937. ''Origins of the Great Wall of China,'' *Geographical Review*, Vol. 27, pp. 529–549.

Lefferts, Walter, 1913. ''The Cattle Industry of the Llanos,'' *Bulletin of the American Geographical Society*, Vol. 45, pp. 180–187.

Leighly, John, 1937. ''Some Comments on Contemporary Geographic Method,'' *Annals of the Association of American Geographers*, Vol. 27, pp. 125–141.

Leighly, John, 1940. ''Environmentalism in the History of Thought,'' *Annals of the Association of American Geographers*, Vol. 30, pp. 64–65.

Leighly, John, 1954. "What Has Happened to Physical Geography?" *Annals of the Association of American Geographers*, Vol. 44, pp. 224–225.

Leighly, John, 1955. "What Has Happened to Physical Geography?" *Annals of the Association of American Geographers*, Vol. 45, pp. 309–318.

Lemert, Ben F., and Rose V. Lemert, 1938. "Popocatepetl," *Journal of Geography*, Vol. 37, pp. 188–195.

Lewis, Peirce F., 1959. "A Cartographic Analysis of Voting Behavior," *Annals of the Association of American Geographers*, Vol. 49, p. 195.

Lewis, Peter W., 1965. "Three Related Problems in the Formulation of Laws in Geography," *Professional Geographer*, Vol. 17, pp. 24–27.

Lewthwaite, Gordon R., 1966. "Environmentalism and Determinism: A Search for Clarification," *Annals of the Association of American Geographers*, Vol. 56, pp. 1–23.

Lewthwaite, Gordon R., 1969. "A Plea for Storage Geography," *Professional Geographer*, Vol. 21, pp. 1–4.

Libbee, Michael, and Thomas J. Wilbanks, 1982. "Program Evaluation as a Strategy for Program Enhancement," *The Professional Geographer*, Vol. 34, pp. 381–386.

Lobeck, A. K., 1942. "Some Geographical Terms and Their English Relatives," *Journal of Geography*, Vol. 41, pp. 130–137.

Lowenthal, David, 1961. "Geography, Experience, and Imagination: Towards a Geographical Epistemology," *Annals of the Association of American Geographers*, Vol. 51, pp. 241–260.

Lukerman, F., 1958. "Toward a More Geographic Economic Geography," *Professional Geographer*, Vol. 10, pp. 2–10.

Lukerman, F., 1960. "On Explanation, Model, and Description," *Professional Geographer*, Vol. 12, pp. 1–2.

Lukerman, Fred, 1960. "The Concept of Location in Classical Geography," *Annals of the Association of American Geographers*, Vol. 50, p. 333.

Lukerman, F., 1962. "Asia, Libya, Europe: Place and Idea," *Annals of the Association of American Geographers*, Vol. 52, p. 348.

Lydolph, Paul E., 1964. "The Russian Sukhovey," *Annals of the Association of American Geographers*, Vol. 54, pp. 291–309.

Marble, Duane F., and John D. Nystuen, 1958. "Commercial Geography of Urban Areas and the Movement of Persons," *Annals of the Association of American Geographers*, Vol. 48, p. 279.

Marcus, Melvin G., 1979. "Coming Full Circle: Physical Geography in the Twentieth Century," *Annals of the Association of American Geographers*, Vol. 69, pp. 521–532.

Marschner, F. J., 1943. "Maps and a Mapping Program for the United States," *Annals of the Association of American Geographers*, Vol. 33, pp. 199–219.

Marston, C. E., 1921. "The Value of Regional Geography," *Journal of Geography*, Vol. 20, pp. 311–314.

Martin, Geoffrey J., 1959. "Political Geography and Geopolitics," *Journal of Geography*, Vol. 58, pp. 441–444.

Martin, Lawrence, 1938. "Who Named Mount Washington?" *Geographical Review*, Vol. 28, pp. 303–305.

Martin, Lawrence, 1950. "William Morris Davis: Investigator, Teacher, and Leader in

Geomorphology,'' *Annals of the Association of American Geographers*, Vol. 40, pp. 172–180.
Mason, Charles H., 1948. ''The Role of the Geographer in Military Planning,'' *Annals of the Association of American Geographers*, Vol. 38, pp. 104–105.
Massip, Salvador, 1932. ''The Geographical Exploration of Cuba by Airplane,'' *Annals of the Association of American Geographers*, Vol. 22, pp. 67–68.
Mather, John R., 1972. ''Applied Climatology—An Old Discipline Faces a Dynamic Future,'' *Professional Geographer*, Vol. 24, pp. 137–141.
Mather, John R., et al., 1980. ''Climatology: The Challenge for the Eighties,'' *Professional Geographer*, Vol. 32, pp. 285–292.
Matley, Ian M., 1966. ''The Marxist Approach to the Geographical Environment,'' *Annals of the Association of American Geographers*, Vol. 56, pp. 97–111.
Maxfield, O. Orland, 1953. ''Principles: An Approach to Economic Geography,'' *Journal of Geography*, Vol. 52, pp. 25–32.
Mayer, Harold M., 1954. ''Geographers in City and Regional Planning,''*Professional Geographer*, Vol. 6, pp. 7–12.
Mayer, Jonathan D., 1982. ''Medical Geography: Some Unsolved Problems,'' *Professional Geographer*, Vol. 34, pp. 261–269.
McAdie, Alexander, 1929. ''The Third Estate of Geography,'' *Annals of the Association of American Geographers*, Vol. 19, pp. 38–39.
McAdie, Alexander, 1931. ''The Commercial Importance of Fog Control,'' *Annals of the Association of American Geographers*, Vol. 21, pp. 91–100.
McCarty, Harold H., 1942. ''A Functional Analysis of Population Distribution,'' *Geographical Review*, Vol. 32, pp. 282–293.
McCarty, H. H., 1953. ''An Approach to a Theory of Economic Geography,'' *Annals of the Association of American Geographers*, Vol. 43, pp. 183–184.
McCarty, Harold H., 1954. ''Land-Use Competition Between Wheat and Corn in the North American Middle West,'' *Annals of the Association of American Geographers*, Vol. 44, p. 227.
McConnell, Wallace R., 1933. ''Problems of Instruction in Geography,'' *Journal of Geography*, Vol 32, pp. 149–151.
McCune, Shannon, 1940. ''Climatic Regions of Tyosen (Korea),'' *Annals of the Association of American Geographers*, Vol. 30, p. 66.
McDonald, James R., 1966. ''The Region: Its Conception, Design, and Limitations,'' *Annals of the Association of American Geographers*, Vol. 56, pp. 516–528.
McMurry, K. C., 1936. ''Geographic Contributions to Land-Use Planning,'' *Annals of the Association of American Geographers*, Vol. 26, pp. 91–98.
McNee, Robert B., 1956. ''Geographic Objectives in General Education: The Regional Method,'' *Journal of Geography*, Vol. 55, pp. 390–394.
McNee, Robert B., 1966. ''The Structure of Geography and Its Potential Contribution to Generalist Education for Planning,'' *Professional Geographer*, Vol. 18, pp. 63–68.
Meinig, D. W., 1972. ''American Wests: Preface to a Geographical Interpretation,'' *Annals of the Association of American Geographers*, Vol. 62, pp. 159–184.
Meir, Avinoam, 1982. ''The Urgency of Teaching History and Philosophy of Geography,'' *Professional Geographer*, Vol. 34, pp. 6–10.
Merriam, Willis B., 1962. ''The Mushroom Industry at Kernett Square, Pennsylvania,'' *Journal of Geography*, Vol. 61, pp. 68–71.

Meyerhoff, Howard A., 1940. "Migration of Erosional Surfaces," *Annals of the Association of American Geographers*, Vol. 30, pp. 66–67.

Mikesell, Marvin W., 1974. "Geography and The Study of Environment: An Assessment of Some Old and New Commitments," in Ian R. Manners and Marvin W. Mikesell, *Perspectives on Environment* (Washington, D.C.: Association of American Geographers, pp. 1–23).

Mikesell, Marvin W., 1978. "Tradition and Innovation in Cultural Geography," *Annals of the Association of American Geographers*, Vol. 68, pp. 1–16.

Miles, Edward J., 1959. "The Geography of Political Attitudes: Some Methodological Considerations," *Annals of the Association of American Geographers*, Vol. 49, pp. 200–201.

Miller, David H., 1965. "Geography, Physical and Unified," *Professional Geographer*, Vol. 17, pp. 1–4.

Miller, George J., 1913. "Some Geographic Influences in the Settlement of Michigan in the Distribution of Its Population," *Bulletin of the American Geographical Society*, Vol. 45, pp. 321–348.

Miller, George J., 1915. "Essentials of Modern Geography and Criteria for Their Determination," *Journal of Geography*, Vol. 13, pp. 129–135.

Miller, Willis H., 1936. "Modern Geography and Current Events," *Journal of Geography*, Vol. 35, pp. 279–284.

Minghi, Julian V., 1963. "Boundary Studies and National Prejudices," *Professional Geographer*, Vol. 15, pp. 4–8.

Mitchell, Adelphia, 1931. "Life Adjustments in Cooks Harbor, Newfoundland," *Journal of Geography*, Vol. 30, pp. 137–144.

Mitchell, Bruce, and Dianne Draper, 1983. "Ethics in Geographical Research," *Professional Geographer*, Vol. 35, pp. 9–17.

Mitchell, Guy Elliott, 1929. "The Topographic Map of the United States," *Economic Geography*, Vol. 5, pp. 382–389.

Monier, Robert B., and Norman E. Green, 1957. "Aerial Photographic Interpretation and the Human Geography of the City," *Professional Geographer*, Vol. 9, pp. 2–5.

Monk, Janice, and Susan Hanson, 1982. "On Not Excluding Half of the Human in Human Geography," *Professional Geographer*, Vol, 34, pp. 11–23.

Moodie, D. W., and John C. Lehr, 1976. "Fact and Theory in Historical Geography," *Professional Geographer*, Vol. 28, pp. 132–135.

Moriarty, Barry M., 1981. "Future Research Directions in American Human Geography," *Professional Geographer*, Vol. 33, pp. 484–488.

Morrill, Richard, 1983. "The Nature, Unity and Value of Geography," *Professional Geographer*, Vol. 35, pp. 1–9.

Morse, Jedidiah, 1825. *Elements of Geography*, New Haven: H. Howe, 6th edition.

Moscheles, Julie, 1937. "The Demographic, Social, and Economic Regions of Greater Prague," *Geographical Review*, Vol. 27, pp. 414–429.

Muehrcke, Phillip C., 1981. "Whatever Happened to Geographic Cartography?" *Professional Geographer*, Vol. 33, pp. 397–405.

Murphey, Rhoads, 1954. "The City as a Center of Change: Western Europe and China, *Annals of the Association of American Geographers*, Vol. 44. pp. 349–362.

Nelson, Helge, 1982. "Geografien som Vetenskap," *Geografiska Annales*, Vol. 64, pp. 119–125.

Newman, James L., 1973. "The Use of the Term 'Hypothesis' in Geography," *Annals of the Association of American Geographers*, Vol. 63, pp. 22–27.

Nunley, Robert E., 1959. "Regional Geography and the Distribution of Population in Costa Rica," *Annals of the Association of American Geographers*, Vol. 49, pp. 204–205.

Olson, Charles E., Jr., 1967. "Accuracy of Land-Use Interpretation from Infrared Imagery in the 4.5 to 5.5 Micron Band," *Annals of the Association of American Geographers*, Vol. 57, pp. 382–388.

Orchard, John E., 1930. "The Introductory Course in Economic Geography in the School of Business," *Journal of Geography*, Vol. 29, pp. 187–198.

Orme, Antony R., 1980. "The Need for Physical Geography," *Professional Geographer*, Vol. 32, pp. 141–148.

Packard, Leonard O., 1913. "Russian Expansion and the Long Struggle for Open Ports," *Journal of Geography*, Vol. 12, pp. 33–39.

Packard, Leonard O., 1924. "The Place of Detail in Geographic Teaching," *Journal of Geography*, Vol. 23, pp. 147–152.

Parkins, A. E., 1931. "The Antebellum South: A Geographer's Interpretation," *Annals of the Association of American Geographers*, Vol. 21, pp. 1–33.

Parkins, A. E., 1934. "The Geography of American Geographers," *Journal of Geography*, Vol. 33, pp. 221–230.

Parró, Alberto Area, 1942. "Census of Peru, 1940," *Geographical Review*, Vol. 32, pp. 1–20.

Parsons, James J., 1959. "Fog Drip from Summer Stratus, with Special Reference to the Berkeley Hills," *Annals of the Association of American Geographers*, Vol. 49, p. 205.

Parsons, James, 1977. "Geography as Exploration and Discovery," *Annals of the Association of American Geographers*, Vol. 67, pp. 1–16.

Pattison, William D., 1960. "On Behalf of the Old Lantern Slides," *The Professional Geographer*, Vol. 22, p. 4.

Pattison, William D., 1962. "High School Geography Project Begins Experimental Year," *Journal of Geography*, Vol. 61, pp. 367–369.

Pattison, William D., 1964. "The Four Traditions of Geography," *Journal of Geography*, Vol. 63, pp. 211–216.

Peattie, Robert, 1940. "The Fields of Environmentalism," *Annals of the Association of American Geographers*, Vol.30, p. 68.

Peattie, Roderick, 1923. "A Function of Modern Geography," *Journal of Geography*, Vol. 22, pp. 279–280.

Peattie, Roderick, 1931. "Height Limits of Mountain Economies," *Geographical Review*, Vol. 21, pp. 415–428.

Pecora, William T., 1969. "New Horizons in Natural Resource Management," *Professional Geographer*, Vol. 21, pp. 73–78.

Peet, Richard, 1975. "The Geography of Crime," *Professional Geographer*, Vol. 27, pp. 277–280.

Peet, Richard, 1975. "Inequality and Poverty: A Marxist-Geographic Theory," *Annals of the Association of American Geographers*, Vol. 65, pp. 564–571.

Philbrick, Allen K., 1958. "Composite Urban Hinterlands," *Annals of the Association of American Geographers*, Vol. 48, p. 285.

Philbrick, Allen K., and Harold M. Mayer, 1951. "A Technique for Visual Examination of Associations of Areal Patterns,'' *Journal of Geography*, Vol. 50, pp. 367–373.

Pickles, John, 1982. '' 'Science' and the Funding of Human Geography,'' *Professional Geographer*, Vol. 34, pp. 387–392.

Picó, Rafael, 1941. ''Geography in American Universities,'' *Journal of Geography*, Vol. 40, pp. 291–301.

Pike, Richard J., 1974. ''Why Not an Extraterrestrial Geography?'' *Professional Geographer*, Vol. 26, pp. 258–261.

Platt, Robert S., 1926. ''Central American Railways and the Pan-American Route,'' *Annals of the Association of American Geographers*, Vol. 16, pp. 12–21.

Platt, Robert S., 1928. ''A Detail of Regional Geography,'' *Annals of the Association of American Geographers*, Vol. 18, pp. 81–126.

Platt, Robert S., 1931. ''Pirovano: Items in the Argentine Pattern of Terrene Occupancy,'' *Annals of the Association of American Geographers*, Vol. 21, pp. 215–237.

Platt, Robert S., 1931. ''An Urban Field Study: Marquette, Michigan,'' *Annals of the Association of American Geographers*, Vol. 21, pp. 52–73.

Platt, Robert S., 1935. ''Field Approach to Regions,'' *Annals of the Association of American Geographers* Vol. 25, pp. 153–172.

Platt, Robert S., 1936. ''A Curacao Farmstead,'' *Journal of Geography*, Vol. 35, pp. 154–156.

Platt, Robert S., 1939. ''Reconnaissance in British Guiana, with Comments on Microgeography,'' *Annals of the Association of American Geographers*, Vol. 29, pp. 105–126.

Platt, Robert S., 1941. ''Latin America in World Affairs,'' *Journal of Geography*, Vol. 40, pp. 321–330.

Platt, Robert S., 1943. ''Regionalism in World Order,'' *Annals of the Association of American Geographers*, Vol. 33, pp. 230–231.

Platt, Robert S., 1946. ''Problems of Our Time,'' *Annals of the Association of American Geographers*, Vol. 36, pp. 1–43.

Platt, Robert S., 1948. ''Determinism in Geography,'' *Annals of the Association of American Geographers*, Vol. 38, pp. 126–132.

Platt, Robert S., 1948. ''Environmentalism Versus Geography,'' *American Journal of Sociology*, Vol. 53, pp. 351–358.

Poole, Sidman P., 1935. ''The Role of Geography in State Planning,'' *Annals of the Association of American Geographers*, Vol. 25, pp. 51–52.

Poole, Sidman P., 1944. ''Geopolitik—Science or Magic,'' *Journal of Geography*, Vol. 43, pp. 1–12.

Posey, C. J., 1920. ''Regional Geography in Minneapolis–St. Paul,'' *Annals of the Association of American Geographers*, Vol. 10, pp. 153–154.

Powers, Pauline Rafter, 1932. ''Teaching North America as a Whole,'' *Journal of Geography*, Vol. 31, pp. 171–173.

Pratt, Joseph Hyde, 1932. ''American Prime Meridians,'' *Geographical Review*, Vol. 32, pp. 233–244.

Preble, Charles S., 1931. ''Visual Aids: Their Use and Abuse,'' *Journal of Geography*, Vol. 30, pp. 120–125.

Pred, Allan, 1984. ''Place as Historically Contingent Process: Structuration and the Time-Geography of Becoming Places,'' *Annals of the Association of American Geographers*, Vol. 74, pp. 279–297.

Price, A. Grenfell, 1934. "White Settlement in Saba Island, Dutch West Indies," *Geographical Review*, Vol. 24, pp. 42–60.

Price, Edward T., 1955. "Values and Concepts in Conservation," *Annals of the Association of American Geographers*, Vol. 45, pp. 64–84.

Prunty, Merle C., 1979. "Clark in the Early 1940s," *Annals of the Association of American Geographers*, Vol. 69, pp. 42–45.

Pryde, Philip R., 1978. "An Alternative Approach to the Applied Geography Interface," *Professional Geographer*, Vol. 30, pp. 1–2.

Quam, Louis O., 1943. "The Use of Maps in Propaganda," *Journal of Geography*, Vol. 42, pp. 21–32.

Raisz, Erwin J., 1931. "The Physiographic Method of Representing Scenery on Maps," *Geographical Review*, Vol. 21, pp. 297–304.

Raisz, Erwin, 1934. "The Rectangular Statistical Cartogram," *Geographical Review*, Vol. 24, pp. 292–296.

Raisz, Erwin, 1936. "Rectangular Statistical Cartograms of the World," *Journal of Geography*, Vol. 35, pp. 8–10.

Raisz, Erwin, 1944. "Our Lopsided Earth," *Journal of Geography*, Vol. 43, pp. 81–91.

Raisz, Erwin, 1946. "Cartography in 1946," *Journal of Geography*, Vol. 45, pp. 347–351.

Raisz, Erwin, 1957. "Geostenography," *Annals of the Association of American Geographers*, Vol. 47, p. 176.

Ratzel, Friedrich, 1923. *Politische Geographie*, Berlin: R. Oldenbourg, 3rd edition.

Raup, H. F., 1936. "Land-Use and Water-Supply Problems in Southern California: Market Gardens of the Palos Verdes Hills," *Geographical Review*,Vol. 26, pp. 264–269.

Raup, Hugh M., 1942. "Trends in the Development of Geographic Botany," *Annals of the Association of American Geographers*, Vol. 32, pp. 319–354.

Reeher, David H., 1958. "Should Our ICBM Launching Site Be Located Outside the U.S.?" *Professional Geographer*, Vol. 10, pp. 8–11.

Reitsma, Hendrik-Jan A., 1982. "Development Geography, Dependency Relations, and the Capitalist Scapegoat," *The Professional Geographer*, Vol. 34, p. 125.

Renner, G. T., Jr., 1926. "Geography's Affiliations," *Journal of Geography*, Vol. 25, pp. 267–272.

Renner, G. T., Jr., 1926. "Some Principles of Commercial Geography," *Journal of Geography*, Vol. 25, pp. 337–342.

Renner, G. T., 1927. "Geographic Elements in the Concept of Heaven," *Annals of the Association of American Geographers*, Vol. 17, pp. 34–35.

Renner, G. T., 1930. "The Geography Curriculum," *Journal of Geography*, Vol. 29, pp. 344–353.

Renner, G. T., Jr., 1931. "The Introductory Geographic Course," *Journal of Geography*, Vol. 30, pp. 33–38.

Renner, G. T., 1935. "The Statistical Approach to Regions," *Annals of the Association of American Geographers*, Vol. 25, pp. 137–145.

Renner, George T., 1944. "What the War Has Taught Us About Geography," *Journal of Geography*, Vol. 43, pp. 321–330.

Renner, George T., 1950. "Some Principles and Laws of Economic Geography," *Journal of Geography*, Vol. 49, pp. 14–22.

Renner, Mayme Pratt, 1929. "Geography in Poetry," *Journal of Geography*, Vol. 28, pp. 292–298.

Richason, Benjamin F., Jr., 1959. "A Geography Classroom in the Sky," *Journal of Geography*, Vol. 58, pp. 317–325.

Richetson, Oliver, Jr., and A. V. Kidder, 1930. "An Archeological Reconnaissance by Air in Central America," *Geographical Review*, Vol. 20, pp. 177–206.

Ridgley, Douglas C., 1925. "Geographic Principles in the Study of Cities," *Journal of Geography*, Vol. 24, pp. 66–78.

Ripley, William Z., 1899. *The Races of Europe*, New York: D. Appleton and Company.

Ristow, Walter W., 1944. "Air Age Geography: A Critical Appraisal and Bibliography," *Journal of Geography*, Vol. 43, pp. 331–343.

Robinson, Arthur H., 1979. "Geography and Cartography Then and Now," *Annals of the Association of American Geographers*, Vol. 69, pp. 97–102.

Robinson, Geoffrey, 1963. "A Consideration of the Relations of Geomorphology and Geography," *Professional Geographer*, Vol. 15, pp. 13–17.

Roepke, Howard G., 1958. "Care and Development of a Wall-Map Collection," *Professional Geographer*, Vol. 10, pp. 11–15.

Roglic, Josip, 1959. "Geographic Thought in Yugoslavia," *Annals of the Association of American Geographers*, pp. 208–209.

Roorbach, G. B., 1914. "The Trend of Modern Geography: A Symposium," *Bulletin of the American Geographical Society*, Vol. 46, pp. 801–816.

Roosevelt, Theodore, 1934. "Land Problems in Puerto Rico and the Philippine Islands," *Geographical Review*, Vol. 24, pp. 182–204.

Rose, John Kerr, 1935. "Corn, Climate, and Statistical Method in the Corn Belt," *Annals of the Association of American Geographers*, Vol. 25, pp. 53–54.

Rose, John Kerr, 1936. "Corn Yield and Climate in the Corn Belt," *Geographical Review*, Vol. 26, pp. 88–102.

Rose, John Kerr, 1954. "Opportunities for Geographers in Government Service," *Professional Geographer*, Vol. 6, pp. 1–6.

Russell, Joseph A., 1939. "Some Applications of Aerial Photography to Geographic Inventory," *Annals of the Association of American Geographers*, Vol. 29, pp. 91–92.

Russell, Joseph A., 1954. "The Theory and Practice of Applied Geography," *Annals of the Association of American Geographers*, Vol. 44, pp. 279–280.

Russell, Richard Joel, 1945. "Post-War Geography," *Journal of Geography*, Vol. 44, pp. 301–312.

Russell, Richard Joel, 1949. "Geographical Geomorphology," *Annals of the Association of American Geographers*, Vol. 39, pp. 1–11.

Sack, Robert David, 1972. "Geography, Geometry, and Explanations," *Annals of the Association of American Geographers*, Vol. 62, pp. 61–78.

Salisbury, Rollin D., 1910. "Physiography in the High School," *Journal of Geography*, Vol. 9, pp. 57–63.

Sauer, C. O., 1915. "Exploration of the Kaiserin Augusta River in New Guinea, 1912–13," *Bulletin of the American Geographical Society*, Vol. 47, pp. 342–345.

Sauer, Carl O., 1921. "Geography as Regional Economics," *Annals of the Association of American Geographers*, Vol. 11, pp. 130–131.

Sauer, Carl O., 1921. "The Problem of Land Classification," *Annals of the Association of American Geographers*, Vol. 11, pp. 3–16.

Sauer, Carl O., 1924. ''The Survey Method in Geography and Its Objectives,'' *Annals of the Association of American Geographers*, Vol. 14, pp. 17–33.

Sauer, Carl O., 1925. *The Morphology of Landscape*, University of California (Berkeley) Publications in Geography, Vol. 2, No. 2, pp. 315–350.

Sauer, Carl O., 1941. ''Foreword to Historical Geography,'' *Annals of the Association of American Geographers*, Vol. 31, pp. 1–24.

Sauer, Carl O., 1941. ''The Personality of Mexico,'' *Geographical Review*, Vol. 31, pp. 353–364.

Sauer, Carl O., 1952. ''Folkways of Social Science,'' *The Social Sciences at Mid-Century: Essays in Honor of Guy Stanton Ford*, Minneapolis: University of Minnesota Press.

Sauer, Carl O., 1956. ''The Education of a Geographer,'' *Annals of the Association of American Geographers*, Vol. 46, pp. 287–299.

Sauer, C. O., 1971. ''The Frontier Years of Ratzel in the United States,'' *Annals of the Association of American Geographers*, Vol. 61, pp. 245–254.

Schaefer, Fred K., 1953. ''Exceptionalism in Geography: A Methodological Examination,'' *Annals of the Association of American Geographers*, Vol. 43, pp. 226–249.

Sears, Paul B., 1958. ''The Inexorable Problem of Space,'' *Science*, Vol. 127, pp. 9–16.

Semple, Ellen Churchill, 1910. ''The Anglo-Saxons of the Kentucky Mountains: A Study in Anthropogeography,'' *Bulletin of the American Geographical Society*, Vol. 42, pp. 561–594.

Semple, Ellen Churchill, 1911 *Influences of Geographic Environment: On the Basis of Ratzel's Systems of Anthropo-Geography*, New York: Henry Holt and Co.

Semple, Ellen Churchill, 1913. ''Japanese Colonial Methods,'' *Bulletin of the American Geographical Society*, Vol. 45, pp. 255–275.

Semple, Ellen Churchill, 1916. ''Pirate Coasts of the Mediterranean,'' *Geographical Review*, Vol. 2, pp. 134–151.

Semple, Ellen Churchill, 1922. ''The Influence of Geographic Conditions Upon Ancient Mediterranean Stock-Raising,'' *Annals of the Association of American Geographers*, Vol. 12, pp. 3–38.

Semple, Ellen Churchill, 1928. ''Ancient Mediterranean Pleasure Gardens,'' *Geographical Review*, Vol. 19, pp. 420–443.

Shaw, Earl B., 1963. ''Geography and Baseball,'' *Journal of Geography*, Vol. 62, pp. 74–76.

Shear, James A., 1965. ''The Long Cold Winter,'' *Annals of the Association of American Geographers*, Vol. 55, p. 646.

Shearer, M. H., 1930. ''Aviators Need Geography,'' *Journal of Geography*, Vol. 29, pp. 371–380.

Shelford, V. E., 1913. ''The Significance of Evaporation in Animal Geography,'' *Annals of the Association of American Geographers*, Vol. 3, pp. 29–42.

Sibley, David, 1981. ''The Notion of Order in Spatial Analysis,'' *Professional Geographer*, Vol. 33, pp. 1–5.

Simonds, Frederick William, 1912. ''Geographic Influences in the Development of Texas,'' *Journal of Geography*, Vol. 10, pp. 277–284.

Smith, Guy-Harold, 1932. ''Cartography in Philately,'' *Annals of the Association of American Geographers*, Vol. 22, p. 76.

Smith, Guy-Harold, 1935. "The Relative Relief of Ohio," *Geographical Review*, Vol. 25, pp. 272–284.

Smith, J. Russell, 1915. "The Desert's Edge," *Bulletin of the American Geographical Society*, Vol. 47, pp. 813–831.

Smith, J. Russell, 1916. "The Island and the Continent at War," *Annals of the Association of American Geographers*, Vol. 6, p. 131.

Smith J. Russell, 1935. "Are We Free to Coin New Terms?" *Annals of the Association of American Geographers*, Vol. 25, pp. 17–22.

Smith, J. Russell, 1939. "The Doomed Valley of the Upper Rio Grande—An Example of Regional Suicide," *Annals of the Association of American Geographers*, Vol. 29, pp. 94–95.

Smith, J. Russell, 1942. "A Note: Suggestions for Illustrating Books," *Annals of the Association of American Geographers*, Vol. 32, pp. 316–317.

Smith, J. Russell, 1943. "Grassland and Farmland as Factors in the Cyclical Development of Eurasian History," *Annals of the Association of American Geographers*, Vol. 33, pp. 135–161.

Smith, J. Russell, 1947. "What Shall the Geography Teacher Teach in the Elementary School? Teach Ideas and Teach Civilization," *Journal of Geography*, Vol. 46, pp. 101–108.

Smith, J. Russell, 1954. "How to Understand a Nation," *Journal of Geography*,Vol. 53, pp. 71–84.

Snyder, David E., 1960. "Peripheral Punch Cards in Geographic Research," *Professional Geographer*, Vol. 12, pp. 13–15.

Soja, Edward W., 1980. "The Socio-Spatial Dialectic," *Annals of the Association of American Geographers*, Vol. 70, pp. 207–225.

Sonnenfield, J., 1960. "Changes in an Eskimo Hunting Technology," *Annals of the Association of American Geographers*, Vol. 50, pp. 172–186.

Spate, O. H. K., 1932. "Factors in the Development of Capital Cities," *Geographical Review*, Vol. 32, pp. 622–631.

Spate, O. H. K., 1960. "Quantity and Quality in Geography," *Annals of the Association of American Geographers*, Vol. 50, pp. 377–394.

Spencer, Joseph Earle, 1935. "Salt in China," *Geographical Review*, Vol. 25, pp. 353–366.

Spencer, J. E., 1941. "Chinese Place Names and the Appreciation of Geographic Realities," *Geographical Review*, Vol. 31, pp. 79–94.

Spencer, J. E., 1958. "An 'Underdeveloped' Population?: The Example of Malaya," *Annals of the Association of American Geographers*, Vol. 48, p. 290.

Spilhaus, Athelstan F., 1942. "Maps of the Whole World Ocean," *Geographical Review*, Vol. 32, pp. 431–435.

Stamp, L. Dudley, 1930. "Burma: An Underdeveloped Monsoon Country," *Geological Review*, Vol. 20, pp. 86–109.

Stamp, L. Dudley, 1934. "Land Utilization Survey as a School and College Exercise," *Journal of Geography*, Vol. 33, pp. 121–130.

Starkey, Otis P., 1950. "A Proposal for a Regional Inventory of Geographic Studies," *Annals of the Association of American Geographers*, Vol. 40, p. 152.

Starkey, Otis P., 1962. "Geography from a Jet," *Journal of Geography*, Vol. 61, pp. 261–266.

Stefansson, V., 1913. "Misconceptions About Life in the Arctic," *Bulletin of the American Geographical Society*, Vol. 45, pp. 17–32.

Stevens, Benj. A., 1914. "Influences of the Mountains of the British Isles Upon Their History," *Journal of Geography*, Vol. 13, pp. 39–45.

Stevens, Benj. A., 1983. "Regionalism in the Great Depression," *Geographical Review*, Vol. 73, pp. 430–446.

Stewart, J. Q., 1947. "Empirical Mathematical Rules Concerning the Distribution and Equilibrium of Population," *Geographical Review*, Vol. 37, pp. 461–485.

Stilgenbauer, Floyd A., 1932. "Cartography and the New Population Map of the United States," *Annals of the Association of American Geographers*, Vol. 22, pp. 77–78.

Stoddart, D. R., 1966. "Darwin's Impact on Geography," *Annals of the Association of American Geographers*, Vol. 56, pp. 683–698.

Stone, Kirk H., 1954. "A Guide to the Investigation and Analysis of Aerial Photos," *Annals of the Association of American Geographers*, Vol. 44, pp. 318–328.

Stone, Kirk H., 1970. "Has the 'Geo' Gone out of Geography?" *Professional Geographer*, Vol. 22, pp. 5–8.

Strahler, Arthur N., 1949. "Recent Developments in Quantitative Analysis of Erosional Landforms," *Annals of the Association of American Geographers*, Vol. 39, p. 65.

Strahler, Arthur N., 1954. "Empirical and Explanatory Methods in Physical Geography," *Professional Geographer*, Vol. 6, pp. 4–8.

Strong, Helen M., 1929. "Geography in Business," *Annals of the Association of American Geographers*, Vol. 19, pp. 48–49.

Strong, Helen M., 1930. "Geography Essential," *Journal of Geography*, Vol. 29, pp. 210–214.

Strong, Helen M., 1937. "Regions of Manufacturing Industry in the United States," *Annals of the Association of American Geographers*, Vol. 27, pp. 23–43.

Sundaram, K. J. G., 1931. "A Deccan Village in India," *Journal of Geography*, Vol. 30, pp. 49–57.

Sutherland, W. J., 1914. "The Vocational Aspects of Regional Geography," *Journal of Geography*, Vol. 12, pp. 308–312.

Sviatlovsky, E. E., and Walter Crosby Eells, 1937. "The Centrogeographical Method and Regional Analysis," *Geographical Review*, Vol. 27, pp. 240–254.

Switzer, J. E., 1936. "Geographical Interdependence," *Journal of Geography*, Vol. 35, pp. 99–105.

Sykes, Godfrey, 1915. "The Isles of California," *Bulletin of the American Geographical Society*, Vol. 47, pp. 745–761.

Taaffe, Edward J., 1974. "The Spatial View in Context," *Annals of the Association of American Geographers*, Vol. 64, pp. 1–16.

Taaffe, Edward J., 1985. "Comments on Regional Geography," *Journal of Geography*, Vol. 84, pp. 96–97.

Taylor, Griffith, 1937. "The Distribution of Pasture in Australia," *Geographical Review*, Vol. 27, pp. 291–294.

Taylor, Griffith, 1939. "Sea to Sahara: Settlement Zones in Eastern Algeria," *Geographical Review*, Vol. 29, pp. 177–195.

Taylor, Griffith, 1942. "Environment, Village and City: A Genetic Approach to Urban

Geography; with Same Reference to Possibilism,'' *Annals of the Association of American Geographers*, Vol. 32, pp. 1–67.
Taylor, Griffith, editor, 1951. *Geography in the Twentieth Century*, New York: Philosophical Library.
Taylor, G. D., and C. M. Matheson, 1958. ''Use of a Photo Copy Machine in Map Reproduction,'' *Professional Geographer*, Vol. 10, pp. 13–14.
Terjung, Werner H., 1976. ''Climatology for Geographers,'' *Annals of the Association of American Geographers*, Vol. 66, pp. 199–222.
Terjung, Werner H., and Stella S-F. Louie, 1973. ''Energy Budget and Photosynthesis of Canopy Leaves,'' *Annals of the Association of American Geographers*, Vol. 63, pp. 109–130.
Terrill, Ross, 1980. *Mao, A Biography*, New York: Harper and Row.
Thoman, Richard S., 1965. ''Some Comments on *The Science of Geography*,'' *Professional Geographer*, Vol. 17, pp. 8–10.
Thomas, Benjamin E., 1956. ''The Legend of Timbuktu,'' *Journal of Geography*, Vol. 55, pp. 434–441.
Thomas, Helen Gross, 1920. ''How Shall We Teach Geography?'' *Journal of Geography*, Vol. 19, pp. 250–254.
Thomas, Katheryne Colvin, 1931. ''Organization of a Unit in Mathematical Geography,'' *Journal of Geography*, Vol. 30, pp. 247–251.
Thompson, Kenneth, 1960. ''Geography—A Problem in Nomenclature,'' *Professional Geographer*, Vol. 12, pp. 4–7.
Thompson, Kenneth, 1969. ''Insalubrious California: Perception and Reality,'' *Annals of the Association of American Geographers*, Vol. 59, pp. 50–64.
Thompson, Will F., 1964. ''How and Why to Distinguish Between Mountains and Hills,'' *Professional Geographer*, Vol. 16, pp. 6–8.
Thornthwaite, C. Warren, 1931. ''The Climates of North America According to a New Classification,'' *Geographical Review*, Vol. 21, pp. 633–655.
Thrower, Norman J. W., and Ronald U. Cooke, 1968. ''Scales for Determining Slope from Topographic Maps,'' *Professional Geographer*, Vol. 20, pp. 181–186.
Tobler, W. R., 1966. ''Medieval Distortions: The Projections of Ancient Maps,'' *Annals of the Association of American Geographers*, Vol. 56, pp. 351–360.
Trewartha, Glenn T., 1926. ''Recent Thought on White Acclimatization in the Wet Tropics,'' *Annals of the Association of American Geographers*, Vol. 16, pp. 36–37.
Trewartha, Glenn T., 1930. ''The Iwaki Basin: Reconnaissance Field Study of a Specialized Apple District in Northern Honshiu, Japan,'' *Annals of the Association of American Geographers*, Vol. 20, pp. 196–223.
Trewartha, Glenn T., 1932. ''The Prairie du Chien Terrace: Geography of a Confluence Site,'' *Annals of the Association of American Geographers*, Vol. 22, pp. 80–81.
Trewartha, Glenn T., 1943. ''The Unincorporated Hamlet: One Element of the American Settlement Fabric,'' *Annals of the Association of American Geographers*, Vol. 33, pp. 32–81.
Trewartha, Glenn T., 1948. ''Some Regional Characteristics of American Farmsteads,'' *Annals of the Association of American Geographers*, Vol. 38, pp. 169–225.
Trewartha, Glenn T., 1953. ''A Case for Population Geography,'' *Annals of the Association of American Geographers*, Vol. 43, pp. 71–97.
Troll, C., 1949. ''Geographic Science in Germany during the Period 1933–1945: A

Critique and Justification,'' *Annals of the Association of American Geographers*, Vol. 39, pp. 99–137.

Tuan, Yi-Fu, 1957. ''Use of Simile and Metaphor in Geographical Descriptions,'' *Professional Geographer*, Vol. 9, pp. 8–11.

Tuan, Yi-Fu, 1964. ''The Problem of Geographical Description,'' *Annals of the Association of American Geographers*, Vol. 54, p. 439.

Tuan, Yi-Fu, 1965. ''Attitudes toward Environment: Themes and Approaches,'' *Annals of the Association of American Geographers*, Vol. 55, pp. 652–653.

Tuan, Yi-Fu, 1975. ''Images and Mental Maps,'' *Annals of the Association of American Geographers*, Vol. 65, pp. 205–213.

Tuan, Yi-Fu, 1976. ''Humanistic Geography,'' *Annals of the Association of American Geographers*, Vol. 66, pp. 266–276.

Tuan, Yi-Fu, 1978. ''Sign and Metaphor,'' *Annals of the Association of American Geographers*, Vol. 68, pp. 363–372.

Tuan, Yi-Fu, 1984. ''Continuity and Discontinuity,'' *Geographical Review*, Vol. 74, pp. 245–256.

Turner, Frederick Jackson, 1914. ''Geographical Influences in American Political History,'' *Bulletin of the American Geographical Society*, Vol. 46, pp. 591–595.

Turner, Frederick Jackson, 1926. ''Geographic Sectionalism in American History,'' *Annals of the Association of American Geographers*, Vol. 16, pp. 35–36.

Ullman, Edward, 1939. ''The Eastern Rhode Island–Massachusetts Boundary Zone,'' *Geographical Review*, Vol. 29, pp. 291–302.

Ullman, Edward L., 1949. ''Mapping the World's Ocean Trade: A Research Proposal,'' *Professional Geographer*, Vol. 1, pp. 19–21.

Ullman, Edward L., 1953. ''Human Geography and Area Research,'' *Annals of the Association of American Geographers*, Vol. 43, pp. 54–66.

Ullman, Edward L., 1954. ''Geography as Spatial Interaction,'' *Annals of the Association of American Geographers*, Vol. 44, pp. 283–284.

Ullman, Edward L., 1958. ''Trade Centers and Tributary Areas of the Philippines,'' *Annals of the Association of American Geographers*, Vol. 48, p. 294.

Ullman, Edward L., and Walter Isard, 1951. ''Toward a More Analytic Economic Geography: The Study of Flow Phenomena,'' *Annals of the Association of American Geographers*, Vol. 41, p. 179.

Ursula, Mary, 1959. ''Changing Patterns of Catholic Population in Eastern United States,'' *Annals of the Association of American Geographers*, Vol. 49, p. 197.

Vale, Thomas R., and Albert J. Parker, 1980. ''Biogeography: Research Opportunities for Geographers,'' *Professional Geographer*, Vol. 32, pp. 149–157.

Van Cleef, Eugene, 1912. ''A Geographic Study of Duluth,'' *Bulletin of the American Geographical Society*, Vol. 44, pp. 401–417.

Van Cleef, Eugene, 1913. ''The Language of Geography,'' *Journal of Geography*, Vol. 11, pp. 235–238.

Van Cleef, Eugene, 1915. ''Geography and the Business Man,'' *Annals of the Association of American Geographers*, Vol. 5, p. 138.

Van Cleef, Eugene, 1915. ''The Sugar Beet in Germany, with Special Attention to Its Relation to Climate,'' *Bulletin of the American Geographical Society*, Vol. 47, pp. 241–258.

Van Cleef, Eugene, 1940. ''The Finns of the Pacific Coast of the United States, and

Consideration of the Problem of Scientific Land Settlement,'' *Annals of the Association of American Geographers*, Vol. 30, pp. 25–38.

Van Cleef, Eugene, 1944. ''Vocational Aspects of Geography,'' *Journal of Geography*, Vol. 43, pp. 241–246.

Van Cleef, Eugene, 1947. ''World Events and Their Implications in the Geography Curriculum,'' *Journal of Geography*, Vol. 46, pp. 91–96.

Van Cleef, Eugene, 1955. ''Must Geographers Apologize,'' *Annals of the Association of American Geographers*, Vol. 45, pp. 105–108.

Van Cleef, Eugene, 1957. ''Some Aspects of Urbanistics,'' *Professional Geographer*, Vol. 9, pp. 2–7.

Van Cleef, Eugene, 1960. ''Geography as an Earth Science,'' *Professional Geographer*, Vol. 12, pp. 8–11.

Van Cleef, Eugene, 1964. ''Confusion or Revolution?'' *Professional Geographer*, Vol. 16, pp. 1–4.

Vinge, Clarence L., 1953. ''The Language of Geography,'' *Professional Geographer*, Vol. 5, pp. 8–9.

Visher, Stephen S., 1930. ''Rainfall and Wind Conditions Retarding Tropical Development,'' *Economic Geography*, Vol. 6, pp. 152–165.

Visher, Stephen S., 1930. ''The Cyclones,'' *Journal of Geography*, Vol. 29, pp. 381–389.

Visher, Stephen Sargent, 1932. ''What Sort of International Boundary Is Best?'' *Journal of Geography*, Vol. 31, pp. 288–296.

Visher, Stephen S., 1934. ''International Boundaries: A Classification and Evaluation,'' *Annals of the Association of American Geographers*, Vol. 24, pp. 71–72.

Visher, Stephen S., 1935. ''Climate Effects of the Proposed Wooded Shelter Belt in the Great Plains,'' *Annals of the Association of American Geographers*, Vol. 25, pp. 63–73.

Visher, Stephen S., 1936. ''The Spread of Population over Europe,'' *Journal of Geography*, Vol. 35, pp. 30–33.

Visher, Stephen S., 1938. ''Influences Locating International Boundaries,'' *Journal of Geography*, Vol. 37, pp. 301–308.

Visher, Stephen S., 1938. ''Rainfall-Intensity Contrasts in Indiana,'' *Geographical Review*, Vol. 28, pp. 627–637.

Visher, Stephen S., 1940. ''Weather Influences on the Yields of Corn, Wheat, Oats and Hay in Indiana, Studied by the Climograph Method,'' *Annals of the Association of American Geographers*, Vol. 30, pp. 77–79.

Visher, Stephen S., 1941. ''Rainfall Conditions as a Southern Handicap,'' *Journal of Geography*, Vol. 40, pp. 302–306.

Visher, Stephen S., 1943. ''Some Regional Contrasts in Precipitation in the United States,'' *Journal of Geography*, Vol. 42, pp. 221–224.

Visher, S.S., 1948. ''Regionalization of Indiana,'' *Annals of the Association of American Geographers*, Vol. 38, pp. 282–300.

Visher, Stephen S., 1952. ''An Aspect of the Social Geography of Indiana,'' *Annals of the Association of American Geographers*, Vol. 42, pp. 98–104.

von Engeln, O. D., 1926. ''A Rejoinder to 'Geography's Affiliations,' '' *Journal of Geography*, Vol. 25, pp. 273–277.

Wagner, Philip L., and Marvin W. Mikesell, eds., 1962. *Readings in Cultural Geography*, Chicago: University of Chicago Press, p. 23.

Walker, Richard A., 1981. "Left-Wing Libertarianism, An Academic Disorder: A Response to David Sibley," *Professional Geographer*, Vol. 33, pp. 5–9.

Ward, Robert De C., 1914. "The Weather Element in American Climates," *Annals of the Association of American Geographers*, Vol. 4, pp. 3–54.

Ward, Robert De C., 1930. "Thirty Thousand Miles of Barograph Curves," *Geographical Review*, Vol. 20, pp. 273–277.

Ward, Robert De C., 1931. "The Literature of Climatology," *Annals of the Association of American Geographers*, Vol. 21, pp. 34–51.

Warman, Henry J., 1944. "Is 'Global Geography' the Answer?" *Journal of Geography*, Vol. 43, pp. 303–306.

Warman, Henry J., 1946. "Geography in a World of Motion," *Journal of Geography*, Vol. 45, pp. 173–178.

Warntz, William, 1957. "Transportation, Social Physics, and the Law of Refraction," *Professional Geographer*, Vol. 9, pp. 2–7.

Watson, J. Wreford, 1953. "Geography in Relation to the Physical and Social Sciences," *Journal of Geography*, Vol. 52, pp. 313–323.

Watson, J. Wreford, 1958. "North America in the Changing World," *Journal of Geography*, Vol. 57, pp. 381–389.

Watson, J. Wreford, 1962. "Geography and History Versus 'Social Studies,' " *Journal of Geography*, Vol. 61, pp. 125–128.

Wheeler, E. P., 2nd, 1935. "The Nain-Okak Section of Labrador," *Geographical Review*, Vol. 25, pp. 240–254.

Wheeler, Lois, 1952. "Experiences in Teaching Third Grade Geography," *Journal of Geography*, Vol. 51, pp. 281–282.

Wheeler, Raymond H., 1939. "History Cycles and Climate," *Annals of the Association of American Geographers*, Vol. 29, pp. 99–100.

Whitaker, J. Russell, 1932. "Regional Interdependence," *Journal of Geography*, Vol. 31, pp. 164–165.

Whitaker, J. R., 1937. "The Study of Cities as a Concluding Unit in Economic Geography," *Journal of Geography*, Vol. 36, pp. 50–54.

Whitaker, J. R., 1940. "Sequence and Equilibrium in Destruction and Conservation of Natural Resources," *Annals of the Association of American Geographers*, Vol. 30, p. 79.

Whitaker, J. R., 1943. "The Place of Geography in the Social Studies: From the Viewpoint of Conservation Education," *Journal of Geography*, Vol. 42, pp. 12–21.

Whitaker, J. R., 1944. "Design for High School Geography," *Journal of Geography*, Vol. 43, pp. 281–296.

Whitaker, J. Russell, 1954. "The Way Lies Open," *Annals of the Association of American Geographers*, Vol. 44, pp. 231–244.

Whitbeck, R. H., 1913. "Economic Aspects of the Glaciation of Wisconsin," *Annals of the Association of American Geographers*, Vol. 3, pp. 62–87.

Whitbeck, R. H., 1914. "A Geographical Study of Nova Scotia," *Bulletin of the American Geographical Society*, Vol. 46, pp. 413–419.

Whitbeck, R. H., 1915. "The St. Lawrence River and Its Part in the Making of Canada," *Bulletin of the American Geographical Society*, Vol. 47, pp. 584–593.

Whitbeck, R. H., 1916. "Geographic Environment as a Factor in Evolution and in Race Differentiation," *Annals of the Association of American Geographers*, Vol. 6, p. 132.

Whitbeck, R. H., 1923. ''Fact and Fiction in Geography by Natural Regions,'' *Journal of Geography*, Vol. 22, pp. 86–94.

Whitbeck, R. H., 1926. ''Adjustments to Environment in South America: An Interplay of Influences,'' *Annals of the Association of American Geographers*, Vol. 16, pp. 1–11.

Whitbeck, R. H., 1926. ''A Science of Geonomics,'' *Annals of the Association of American Geographers*, Vol. 16, pp. 117–123.

Whitbeck, R. H., 1932. ''The Agricultural Geography of Jamaica,'' *Annals of the Association of American Geographers*, Vol. 22, pp. 13–27.

Whitbeck, R. H., 1933. ''The Lesser Antilles—Past and Present,'' *Annals of the Association of American Geographers*, Vol. 23, pp. 21–26.

Whitbeck, R. H., 1934. ''Sources of New Terms,'' *Annals of the Association of American Geographers*, Vol. 24, pp. 87–88.

White, Gilbert F., 1958. ''Introductory Graduate Work for Geographers,'' *Professional Geographer*, Vol. 10, pp. 6–8.

White, Gilbert F., 1962. ''Critical Issues Concerning Geography in the Public Service,'' *Annals of the Association of American Geographers*, Vol. 52, pp. 279–280.

White, Gilbert F., 1985. ''Geographers in a Perilously Changing World,'' *Annals of the Association of American Geographers*, Vol. 75, pp. 10–16.

White, Langdon, 1928. ''The Iron and Steel Industry of the Birmingham, Alabama, District,'' *Economic Geography*, Vol. 4, pp. 349–365.

Whittlesey, D. S., 1925. ''A Tentative Map of Geographic Regions of Africa,'' *Annals of the Association of American Geographers*, Vol. 25, p. 49.

Whittlesey, D. S., 1927. ''Devices for Accumulating Geographic Data in the Field,'' *Annals of the Association of American Geographers*, Vol. 17, pp. 72–78.

Whittlesey, Derwent, 1929. ''Sequent Occupance,'' *Annals of the Association of American Geographers*, Vol. 19, pp 162–165.

Whittlesey, Derwent, 1931. ''The Urbanization of a Farm Village,'' *Annals of the Association of American Geographers*, Vol. 21, pp. 142–143.

Whittlesey, Derwent, 1932. ''Trans-Pyrenean Spain,'' *Annals of the Association of American Geographers*, Vol. 22, p. 84.

Whittlesey, Derwent, 1936. ''Major Agricultural Regions of the Earth,'' *Annals of the Association of American Geographers*, Vol. 26, pp. 199–240.

Whittlesey, Derwent, 1937. ''New England,'' *Annals of the Association of American Geographers*, Vol. 27, pp. 168–170.

Whittlesey, Derwent, 1942. ''Geography and Politics,'' *Annals of the Association of American Geographers*, Vol. 32, pp. 142–143.

Whittlesey, Derwent, 1943. ''Geopolitics, a Program for Action,'' *Annals of the Association of American Geographers*, Vol. 33, pp. 97–98.

Whittlesey, Derwent, 1943. ''The Place of Geography in the Social Studies for Orientation in a World of Changing National Boundaries,'' *Journal of Geography*, Vol. 42, pp. 1–6.

Whittlesey, Derwent, 1945. ''The Horizon of Geography,'' *Annals of the Association of American Geographers*, Vol. 35, pp. 1–36.

Whittlesey, D., 1954. ''The Regional Concept and the Regional Method,'' in P. E. James, and C. F. Jones, eds., *American Geography, Inventory and Prospect*, Syracuse: Syracuse University Press, pp. 19–68.

Wigmore, John H., 1929. "A Map of the World's Law," *Geographical Review*, Vol. 19, pp. 114–120.
Wilbanks, Thomas J., 1985. "Geography and Public Policy at the National Scale," *Annals of the Association of American Geographers*, Vol. 75, p. 7.
Wilhelm, E. J., Jr., 1968. "Biogeography and Environmental Science," *Professional Geographer*, Vol. 20, pp. 123–125.
William-Olsson, W., 1940. "Stockholm: Its Structure and Development," *Geographical Review*, Vol. 30, pp. 420–438.
Williams, Charles Edwin, 1965. "A Map of the Wild Fur Harvest in Anglo America," *Annals of the Association of American Geographers*, Vol. 55, pp. 656–657.
Williams, Frank E., 1932. "Crossing the Andes at 41°S," *Annals of the Association of American Geographers*, Vol. 22, p. 85.
Wilson, Henry, 1918. "The Geography of Culture and the Culture of Geography," *Geographical Teacher*, Vol. 9, p. 196.
Wolfanger, Louis A., 1931. "A Single Key to Soil Geography," *Journal of Geography*, Vol. 30, pp. 330–338.
Wolpert, Julian, 1976. "Opening Closed Spaces," *Annals of the Association of American Geographers*, Vol. 66, pp. 1–13.
Wood, Walter F., 1954. "The Dot Planimeter, A New Way to Measure Map Area," *Professional Geographer*, Vol. 6, pp. 12–14.
Wright, John K., 1925. "The History of Geography: A Point of View," *Annals of the Association of American Geographers*, Vol. 15, pp. 192–201.
Wright, John K., 1932. "Sections and National Growth," *Geographical Review*, Vol. 22, pp. 353–360.
Wright, John K., 1936. "The Diversity of New York City," *Geographical Review*, Vol. 26, pp. 620–629.
Wright, John K., 1942. "Map Makers Are Human," *Geographical Review*, Vol. 32, pp. 527–544.
Wright, John K., 1944. "Training for Research in Political Geography," *Annals of the Association of American Geographers*, Vol. 34, pp. 190–201.
Wright, John K., 1947. "Terrae Incognitae: The Place of the Imagination in Geography," *Annals of the Association of American Geographers*, Vol. 37, pp. 1–15.
Wright, John K., 1963. "Wild Geographers I Have Known," *Professional Geographer*, Vol. 15, pp. 1–4.
Zelinsky, Wilbur, 1958. "Population Geography and the Problems of Underdevelopment," *Annals of the Association of American Geographers*, Vol. 48, p. 229.
Zelinsky, Wilbur, 1975. "The Demigod's Dilemma," *Annals of the Association of American Geographers*, Vol. 65, pp. 123–143.
Zobler, Leonard, 1958. "Decision Making in Regional Construction," *Annals of the Association of American Geographers*, Vol. 48, pp. 140–148.

# INDEX

**About the Compilers**

JAMES O. WHEELER is Professor of Geography at the University of Georgia, Athens. He is the author of *Economic Geography*, *American Metropolitan Systems*, and over seventy articles.

FRANCIS M. SIBLEY is Associate Professor of English in the Division of Arts and Sciences at Urbana University in Ohio. He has contributed to *Southern Humanities Review* and *The American Scholar*.